Sebastian Probst

Hybrid quantum system based on rare earth doped crystals

Experimental Condensed Matter Physics
Band 16

Herausgeber
Physikalisches Institut

Prof. Dr. Hilbert von Löhneysen
Prof. Dr. Alexey Ustinov
Prof. Dr. Georg Weiß
Prof. Dr. Wulf Wulfhekel

Eine Übersicht aller bisher in dieser Schriftenreihe erschienenen Bände finden Sie am Ende des Buchs.

Hybrid quantum system based on rare earth doped crystals

by
Sebastian Probst

Scientific
Publishing

Dissertation, Karlsruher Institut für Technologie (KIT)
Fakultät für Physik
Tag der mündlichen Prüfung: 23. Januar 2015
Referenten: Prof. Dr. Alexey V. Ustinov, Prof. Dr. Pavel A. Bushev

Impressum

KIT Scientific Publishing

Karlsruher Institut für Technologie (KIT)
KIT Scientific Publishing
Straße am Forum 2
D-76131 Karlsruhe

KIT Scientific Publishing is a registered trademark of Karlsruhe Institute of Technology. Reprint using the book cover is not allowed.

www.ksp.kit.edu

Print on Demand 2016

ISSN 2191-9925
ISBN 978-3-7315-0345-3
DOI 10.5445/KSP/1000045903

Contents

1 Introduction

In 2012, David Wineland and Serge Haroche were awarded the Nobel prize in physics ''for ground-breaking experimental methods that enable measuring and manipulation of individual quantum systems''. Wineland's experiments demonstrated the coherent control of individual trapped ions, whereas Haroche managed to control and manipulate microwave photons in a cavity using Rydberg atoms. Only two years later, it is possible to control and manipulate quantum systems consisting of multiple atoms. This opens up new prospects for those systems serving as prototype quantum computers and quantum simulators. One day, these machines may efficiently solve some of today's hard computational problems.

In order to be able to appreciate these achievements, we have to ask: What makes quantum systems so special? The answer to this question is founded on the fundamental difference between classical and quantum information. The smallest unit of classical information as we use it in our computers and cell phones today is represented by 0 and 1, and is called a *bit*. In contrast, quantum information is represented by a *qubit*, which has two so-called eigenstates $|0\rangle$ and $|1\rangle$. The difference to a classical bit is that quantum information is described by a wave function $|\Psi\rangle$, which can be in an arbitrary normalized superposition of these two states

$$|\Psi\rangle = \sin(\theta/2)\,|0\rangle + \cos(\theta/2)e^{i\phi}\,|1\rangle\ . \tag{1.1}$$

Here, ϕ and θ represent arbitrary angles. A familiar object represented by two angles as degrees of freedom is the surface of a sphere. This implies that the qubit can be in one of an infinite amount of different states on this surface whereas the classical bit can only be in one of two states, i.e. the north or south pole. However, once we attempt to retrieve information about the qubit state, it turns out that we either get $|0\rangle$ or $|1\rangle$ (with a certain probability depending on the actual qubit state $|\Psi\rangle$). In other words, in each measurement, all the information collapses on one of the two eigenstates. Only with a repeated measurement of an infinite number of identical copies of the state one could infer the actual quantum state. So, it seems that there is no advantage in using a qubit over a classical bit. However, the power of quantum information emerges when combining a lot of qubits. As long as the qubits are not measured, they interact with each other taking into account their full information. To cut a long story short, this enables very fast parallel (quantum) computing. Any classical algorithm can run on such a quantum computer. Additionally, a certain class of problems can be solved exponentially faster than on any computer working with classical bits. That is why a lot of research is invested in building such a machine.

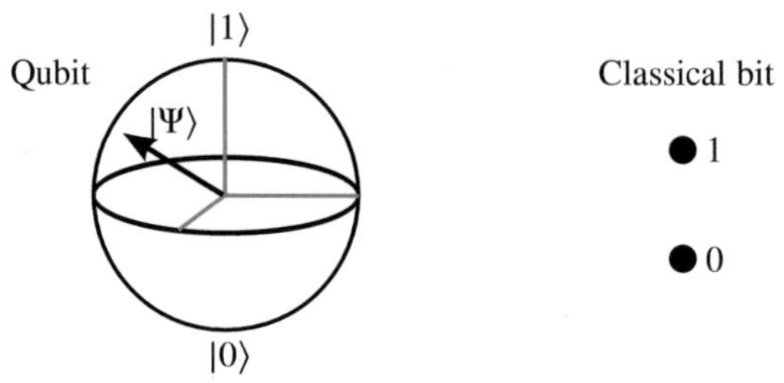

Figure 1.1: Qubit vs. classical bit.

An important resource of quantum information is called *entanglement*. Imagine two qubits A and B described by the common wave function

$$|\Psi\rangle = \frac{1}{\sqrt{2}} \left(\left|0^A 0^B\right\rangle + \left|1^A 1^B\right\rangle \right) . \tag{1.2}$$

Now, we separate the qubits far away from each other and measure qubit A. By measuring A, the qubit state collapses either on $\left|0^A\right\rangle$ or $\left|1^A\right\rangle$. Intriguingly, this measurement determines the outcome of the measurement of qubit B. Two parties sharing such an entangled state can use its properties to communicate in an absolute secure way.

In order to enable quantum communication over large distances, one requires a network distributing quantum states [Cir+97; Kim08]. D. P. DiVincenzo has come up with two criteria, which any future quantum network should fulfill [DiV00].

1. *The ability to interconvert stationary and flying qubits.*
2. *The ability to faithfully transmit flying qubits between specified locations.*

Photons are well suited to transmit quantum information over long distances. They can be guided in optical fibers with very low attenuation on the order of 0.17 dB/km at 1550 µm wavelength[1]. In classical optical fiber communication, amplifiers are used to compensate for the photon losses in the fiber. This is no longer possible in the quantum regime because a quantum state cannot be amplified without destroying part of its information. Instead, *quantum repeaters* are used to bridge even longer distances [GT07]. These employ additional photons to distribute entanglement between the sender, the repeater stations and the recipient such that the message qubit gets teleported from the sender to the recipient. In this way, the photon carrying the quantum information does not physically travel all the way to the recipient.

For a quantum repeater to work reliably, it needs to be able to store quantum information for some time and retrieve it on-demand [Sim+10]. Therefore, Quantum memories are required. They can be implemented in a variety of systems, for instance single trapped ions [Bli+04], atoms [Ros+07], single spins [Mau+12], two-level defects [Nee+08] and spin ensembles [Ste+12]. At the nodes of a future quantum network the flying qubits (photons) are converted

[1] `http://www.corning.com/opticalfiber/products/SMF-28_ULL_fiber.aspx`

into stationary qubits and the information is then processed in quantum computers [OBr+14; McA+12]. Again, such converters may be realized in a variety of different physical systems. Recently, a basic quantum network connecting two single atoms has been demonstrated [Rit+12; Hof+12].

A crucial requirement for the practical application of any physical realization of a quantum system is scalability. This means that there are no constraints, which would limit the number of physical qubits in a system. In terms of scalability and fast operation times, solid state systems such as superconducting (SC) quantum circuits [CW08], nanomechanical devices [OCo+10] and spin doped solids [Wu+10] are well suited to perform local computations. Typically these devices operate at microwave or radio frequencies and in a cryogenic environment in order to eliminate thermal noise. However, it is technically very difficult to build long and cryogenic microwave cables. Thus, this frequency regime is less suited for long distance quantum communication. In order to realize a remote connection between such systems, a coherent quantum converter is needed, which can reversibly convert the microwave or RF signal to an optical photon [Tog+10; Sag+11; Cla+11; De +12]. Based on nanomechanical resonators, a prototype device operating in the classical regime has been reported recently [And+14].

SC quantum circuits operate in the microwave frequency range between 1 and 10 GHz. Instead of natural atoms, macroscopic Josephson qubits are used, which are basically anharmonic electromagnetic oscillators. The field has made tremendous progress in the recent years. This includes the creation of arbitrary quantum states [Hof+09], the demonstration of an elementary quantum processor, which could run Shor's algorithm for prime factorization [Luc+12], and the implementation of a quantum error correction algorithm [Ree+12]. The SC quantum circuit architecture is scalable but suffers from qubits with short coherence times. The coherence time determines how long a quantum state can be stored until it decays. A drastic improvement came along by introducing a new architecture based on Josephson qubits in a three-dimensional cavity [Pai+11; Rig+12]. In contrast, natural systems such as atoms or nuclear spins, feature much longer coherence times. This comes with the price that they are much harder to control and manipulate.

So, each physical quantum system has its advantages and drawbacks such that the way towards a successful quantum technology may lie in the coherent integration of different quantum systems. Due to their good performance and scalability, hybrid systems based on SC coplanar waveguide (CPW) resonators coupled to polar molecules [Rab+06; TNM08], trapped cold atoms [Ver+09; Pet+09] or to spins in solids [Ima09] have been proposed in the years from 2006 to 2009. Whereas the first two proposals are technically extremely challenging due to the need for ultra high vacuum and traps above a superconducting chip, the latter idea has been proven to be quite fruitful. Today, the research activity is mainly focused on circuit and cavity QED experiments with spin ensembles consisting of NV-centers in diamonds [Sch+10; Ams+11; Kub+12; San+12; Gre+14; Put+14]. Strong coherent coupling between NVs and a SC resonator as well as with a SC qubit has been demonstrated [Kub+10; Zhu+11; Kub+11]. The first attempts to couple iron group impurity spins to superconducting resonators yielded only weak coupling [Sch+10]. Recently, coherent coupling to iron group ions could be demonstrated by employing a whispering gallery mode 3D resonator [Gor+14a]. Other types of spin ensembles under investigation include phosphor [Tyr+12] and bismuth donors [Geo+10] in silicon bearing the potential of coherence times on the order of a second. It should be noted that strong coupling of three dimensional cavities to organic molecule spin ensembles has also been demonstrated in conventional room temperature electron spin resonance (ESR) experiments [Chi+10; Abe+11]. All spin ensembles mentioned so far were paramagnetic ensembles but recently, strong coupling of three dimensional cavities to ferromagnetic ensembles has been reported as well [Zha+14; Tab+14; Gor+14b].

An interesting alternative presents itself by rare earth elements due to their large spin tuning rate [KC80] and long optical and spin coherence times [Böt+06a; Ber+07; Rak+09]. In particular, the rare earth elements neodymium and erbium additionally possess optical transitions in standard telecommunication bands. This makes them very attractive for a microwave to optical quantum media converter. Tremendous progress has been achieved with optical quantum memories based on rare earth elements such as multimode storage [Usm+10] and the implementation of a teleportation scheme [Bus+14]. Additionally, in the microwave regime, ensembles of erbium spins coupled to SC CPW resonators have been studied before [Bus+11; Sta+12]. However, at that time, the strong coupling regime was inhibited by the large inhomogeneous spin linewidth. During this thesis we were able to demonstrate coherent strong coupling of erbium spin ensembles to superconducting as well as non-superconducting resonators [Pro+13; Tka+14; Pro+14c]. This paves the way towards the realization of a rare earth based reversible microwave to optical quantum converter [OBr+14].

In this thesis, we study hybrid systems based on the rare earth elements erbium (Er) and neodymium (Nd) doped into Y_2SiO_5 (YSO) and $YAlO_3$ (YAlO) crystals. The focus lies on Er due to its optical transition in the telecom C-band at 1.54 µm making it an obvious choice as core element of a reversible optical to microwave quantum converter. In this work, we investigate the physics of spin ensembles coupled to quantum circuits, and demonstrate experimentally the potential of the rare earth approach for application as coherent quantum memory and reversible quantum converter.

The thesis commences with an introduction into paramagnetic impurities in crystals and their spectroscopic properties. Then, the theoretical concepts of hybrid quantum systems are presented. The theory part concludes with the description of the experimental setup.
The experimental part of this thesis is divided into three chapters. First, the continuous wave (cw) microwave spectroscopy of the rare earth samples coupled to SC resonators is presented. The magnetic anisotropy is investigated and we demonstrate strong coupling to both Er in YSO and YAlO and Nd in YAlO. Moreover, we study the AC g-factor of Er:YSO with a three dimensional cavity operating in the strong coupling regime. In order to integrate rare earth crystals with SC circuits on an application level, it is important to locally embed rare earth based memory elements on the SC chip. In collaboration with N. Kuckharchyk and Prof. A. D. Wieck from the RUB Bochum, samples of focused ion beam implanted Er:YSO have been investigated with our on-chip ESR setup.

Part 2 discusses the time resolved microwave spectroscopy of a 50 ppm doped Er:YSO spin ensemble strongly coupled to a copper CPW resonator. The chapter focuses on the coherence properties of the Er spin ensemble and presents our results on the storage of multiple coherent microwave pulses.

Finally, part 3 concentrates on the optical transition of the erbium spin ensemble. Here, the optical properties at milli-Kelvin temperatures are studied with particular focus on the interplay between microwave pulses and optical transmission. This is a prerequisite for the implementation of a coherent reversible quantum converter.

2 Paramagnetic Ions in Solids

This chapter provides a general introduction into the properties of paramagnetic impurities in crystals and the theoretical background for their investigation with electron spin resonance and optical spectroscopy.

2.1 Transition ions of the 4f group

The rare earth (RE) group or lanthanide series of elements, named after their first element Lanthanum, consists of 15 elements which all have very similar chemical and physical properties. For instance, they are all metals having a silvery, shiny surface on fresh cut edges oxidizing quickly in moist air. The configuration of the outer orbitals of all RE elements is $5s^2 5p^6 5d^1 6s^2$ and along the series the underlying 4f shell gets filled. Since the outer shells dominate the chemical und physical properties of an element, the 4f elements all behave similarly. Typically, RE elements appear as tripositive ions in many compounds. In this thesis, we focus on the crystals Y_2SiO_5 and $YAlO_3$, where the Yttrium (Y^{3+}) can be replaced by a RE^{3+} ion. As schematically shown in Fig. 2.1, the 4f shell of the tripositive ion is well shielded from the environment by the $5s^2$ and $5p^6$ shells, which is the reason for its sharp lines in optical spectra and good coherence properties at low temperatures [AB12; Böt+06b; Böt+06a]. Among all RE elements every second ion namely Ce^{3+}, Nd^{3+}, Sm^{3+}, Gd^{3+}, Dy^{3+}, Er^{3+} and Yb^{3+} is a so called Kramers ion and possesses only a twofold degenerate spin groundstate, which can be accessed by microwave spectroscopy after lifting the degeneracy by a magnetic field. Therefore, these elements are well suited for the coupling to superconducting circuits, which operate in the microwave frequency range. All RE elements offer optical[1] transitions, but only two of the Kramers ions (Nd^{3+}, Er^{3+}) offer optical transitions in standard telecommunication bands of fiber-optic communication. These bands are defined by distinct parts of the absorption spectrum of silica glass optical fibers with minimum absorption and negligible dispersion. This makes Er and Nd excellent candidates for application in quantum communication because they can be both interfaced with superconducting qubit processors and the optical communication band at 1.54 µm (Er) and 1.30 µm (Nd), respectively [Bus+11].

2.2 Rare earth impurities in crystal fields

In the experiments presented in this thesis, we used rare earth doped Yttrium Orthosilicate (Y_2SiO_5) and Yttrium Orthoaluminate ($YAlO_3$), where the rare earth element replaces the Yttrium in the crystal.

[1] Here and throughout the thesis, the word "optical" includes visible as well as near infrared wavelengths.

Y_2SiO_5 (YSO) belongs to the C^6_{2h} space group in Schönflies notation, meaning that it is a monoclinic crystal with a two-fold rotational axis and a mirror plane perpendicular to the rotational axis denoted by b. The dimensions of the unit cell are $a = 1.041$ nm, $b = 0.6721$ nm and $c = 1.249$ nm, and the angle between a and c is 102°39′ [Sun+08]. There is another useful coordinate system (b, D_1, D_2), which is based on the optical extinction axis D_1 and D_2 [LWM92]. D_1 is oriented 23.8° from the c-axis and 78.7° from the a-axis, and D_2 is perpendicular to D_1. In this thesis, the direction of an applied magnetic field is given by the angles θ and ϕ, which are defined with respect to the b and D_1 axes. Typically, this coordinate system is used to describe the magnetic anisotropy of RE doped YSO, which will be discussed later.

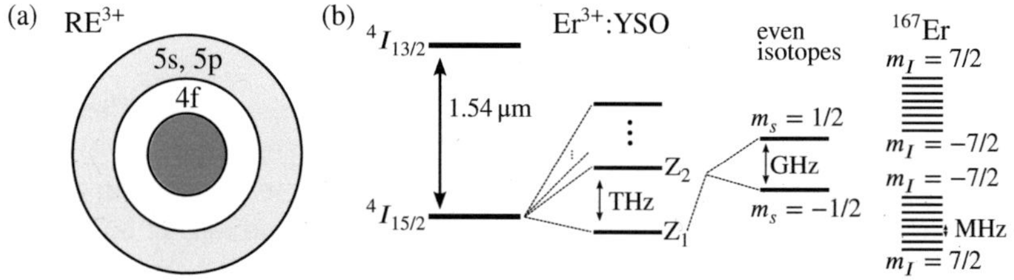

Figure 2.1: (a) Schematic structure of a tripositive rare earth ion: The 4f shell is shielded by the closed 5s and 5p shells. (b) Level structure of Er^{3+} doped into Y_2SiO_5.

Figure 2.1(b) shows the level scheme of Er doped into YSO. The erbium ion has an optical transition in the telecom C-band at 1.54 µm from the $^{2S+1}L_J =^4 I_{15/2}$ to the $^4I_{13/2}$ state. Here, S, L and J denote the spin, orbital and total magnetic moment, respectively. The host crystal YSO perturbs the $^4I_{15/2}$ state by a weak crystal field and therefore, it splits up into eight $(J + 1/2)$ Kramers doublets Z_n with transition frequencies in the THz range [Dou+95]. Since all experiments are carried out at cryogenic temperatures, only the lowest doublet is populated and one can truncate the system to an effective spin 1/2 system. The degeneracy of this spin level can be lifted by applying a magnetic DC field, and the resulting Zeeman shift yields a splitting in the GHz frequency range at moderate fields on the order of 100 mT. Only the odd isotope ^{167}Er with a natural abundance of 22.9% features an additional nuclear spin of 7/2 and effects a hyperfine splitting such that the spin 1/2 transition splits up into eight allowed transitions, see Fig. 2.1(b) [Bus+11]. The magnetic properties of the effective spin 1/2 system in the Er^{3+} ion are determined by the large unquenched total orbital moment J in the crystal field yielding a large magnetic moment of almost 7 μ_B and a large magnetic anisotropy due to the low axial symmetry C_{2h}, which is expressed in terms of a **g**-tensor. The full effective Hamiltonian is given by

$$H = \mu_B \mathbf{B} \cdot \mathbf{g} \cdot \mathbf{S} + \mathbf{I} \cdot \mathbf{A} \cdot \mathbf{S} + \mathbf{I} \cdot \mathbf{Q} \cdot \mathbf{I} - \mu_n g_n \mathbf{B} \cdot \mathbf{I}. \tag{2.1}$$

B denotes the applied magnetic field, **S** is the electronic spin, **I** is the nuclear spin, **A** and **Q** are the hyperfine and quadrupol tensors, respectively, and μ_n is the nuclear magnetic moment with its nuclear g-factor g_n [Gui+06], [AB12] (chapter 3). The first term describes the Zeeman interaction, the second and third the hyperfine and quadrupol interaction, respectively, and the last term represents the nuclear Zeeman interaction. For the most abundant even Erbium isotopes, only the first term of Eq. (2.1) remains.

In the Y_2SiO_5 crystal, the erbium can replace the yttrium ions at two different positions with either 6 or 7 oxygen neighbor atoms. These two nonequivalent crystallographic sites have local C_1 symmetry and their optical transitions $^4I_{15/2} \rightarrow {}^4I_{13/2}$ have slightly different wavelengths. Site 1 has a vacuum transition wavelength of 1536.48 nm and site 2 has 1538.90 nm at zero magnetic field [Böt+06b]. Each of the sites is divided into two magnetic classes due to the C_{2h} symmetry such that one has to use 4 different g-tensors in order to describe the spectrum [KC80; Gui+06].

Rare earth ions in $YAlO_3$ (YAlO), behave to a certain degree similar to RE ions in YSO and the level scheme shown in Fig. 2.1 is comparable. Here, only the differences will be highlighted. YAlO belongs to the D^16_{2h} space group and is an orthorhombic crystal, which shows a weaker magnetic anisotropy than YSO [ARM97]. The unit cell of YAlO consists of four magnetic inequivalent classes for the RE dopant ions, which corresponds to the four distorted perovskite cells of the unit cell. However, this number is reduced. Because of the crystal's symmetry, they subside to two effective magnetic classes. In particular, it is the mirror symmetry of the yttrium ions in the (001) plane and the inversion symmetry through the aluminum sites [Bri+09].

2.3 Electron spin resonance spectroscopy

The following sections are based on the book by A. Abragam and B. Bleaney [AB12], and the book by A. Schweiger and G. Jeschke [SE01].

2.3.1 Electronic and nuclear moments

A paramagnetic substance has no net magnetic moment in the absence of a magnetic field. However, if an external magnetic field is applied, it acquires a magnetic moment, which is proportional and parallel to the external magnetic field. Such a material consists of magnetic dipoles, which, in zero magnetic field, are randomly oriented. These magnetic dipoles can only occur if the atoms or ions possess a net angular momentum **G**, which is related to the magnetic moment $\boldsymbol{\mu}$ of such a dipole by the gyromagnetic ratio γ, $\boldsymbol{\mu} = \gamma \mathbf{G}$. The (classical) equations of motion of such a dipole in a magnetic field **B** are given by

$$\frac{\mathrm{d}\boldsymbol{\mu}}{\mathrm{d}t} = \gamma \boldsymbol{\mu} \times \mathbf{B} . \tag{2.2}$$

One can show, that the magnetic moment $\boldsymbol{\mu}$ precesses around the direction of the magnetic field with an angular frequency of $\boldsymbol{\omega}_\mathrm{L} = -\gamma \mathbf{B}$, which is called Larmor frequency. The component of $\boldsymbol{\mu}$ parallel to $\mathbf{B}$ remains constant so that the so-called Zeeman energy of the dipole in the magnetic field remains fixed, too,

$$E = -\boldsymbol{\mu} \cdot \mathbf{B} . \tag{2.3}$$

Consider a free ion, which has a net angular momentum. Then, the electron system of that ion will possess a permanent magnetic moment and the gyromagnetic ratio reads as $\gamma = -g(e/2mc)$. The minus sign comes from the negative charge e of the electron with mass m. The quantity g depends on the contributions of spin $\mathbf{S}$ and orbital momentum $\mathbf{L}$ to the total angular momentum. If only the orbital momentum is present, $g = g_L = 1$, and for only spin contributions, $g = g_S = 2$. Here, we neglect the QED corrections to these quantities, which are on the order of 10^{-4}. For the general case of spin-orbit coupling $\mathbf{J} = \mathbf{L} + \mathbf{S}$, the value of g is given by the Landé formula

$$g_J = \frac{3}{2} - \frac{L(L+1) - S(S+1)}{2J(J+1)} . \tag{2.4}$$

The total magnetic moment is given by $\boldsymbol{\mu} = -g_J \mu_B \mathbf{J}$, where $\mu_B = e\hbar/2mc$ denotes the Bohr magneton. The same considerations also hold for nuclear magnetic moments but instead of μ_B, the nuclear magneton $\mu_n = e\hbar/2Mc \sim 10^{-3}\mu_B$ has to be employed where M denotes the mass of the proton. An alternative way of writing this is to define $g_I = g_J(m/M)$ and keep μ_B in the definition of the total magnetic moment.

If the free ion possesses a nuclear angular momentum $\mathbf{I}$, it is typically coupled to $\mathbf{J}$ by the following Hamiltonian

$$H = a(\mathbf{J} \cdot \mathbf{I}) . \tag{2.5}$$

This forms a number of energy levels with quantum numbers $F = (J+I), (J+I-1), ..., |J-1|$ with a separation of $\Delta E = aF$. The Zeeman splitting of such a level for small magnetic fields with respect to the hyperfine interaction energy is given by $\mu_F = -g_F \mu_B F$, where

$$g_F = \frac{F(F+1)\left(g_J - g_I\right) + [J(J+1) - I(I+1)]\left(g_J - g_I\right)}{2F(F+1)} . \tag{2.6}$$

However, in typical ESR experiments, the hyperfine energy is usually smaller than the Zeeman energy. In the strong field limit, the contribution of the nucleus to the energy spectrum is treated as a perturbation of the external magnetic field and the total energy spectrum is given by

$$E = g_J \mu_B J_z B + a J_z I_z - g_I \mu_B I_z B . \tag{2.7}$$

The subscript z denotes the z-component of the vector.

2.3.2 Magnetic resonance

A magnetic dipole precesses in a static magnetic field with an angular frequency $\omega_L = -\gamma B$. This effects an oscillating magnetic field normal to the static magnetic field and the magnetic moment interacts with an external oscillatory field $B_1 \cos(\omega t)$, which is perpendicular to the static magnetic field. If the frequency of the external AC field matches the precession frequency $\omega = \omega_L$, it is possible to manipulate the component pointing parallel to the static field $\mu \cos\alpha$ even if $H_1 \ll H$. (A full quantum mechanical description will be given in Sec. 3.2.2.) Changing the component of μ parallel to $\mathbf{B}$ will change the energy of the dipole

$$E = -\mu B \cos\alpha = -\gamma G H \cos\alpha\,. \tag{2.8}$$

Since the angular momentum is quantized, $G\cos\alpha$ can only take values $\hbar M$, $\hbar(M-1)$, ... The smallest change of momentum is $\Delta M = \pm 1$ requiring one quantum of energy $\hbar\omega$. Hence, electron spin resonance spectroscopy allows to determine the value of the g-factor g by measuring the resonance frequency for a given magnetic field

$$g = \frac{\mu_B B}{\hbar\omega}\,. \tag{2.9}$$

Classical dynamics of the driven spin

The equations of motion of the classical magnetic moment (see Eq. (2.2)) already provide a good understanding of the evolution of a spin which is subject to a resonant driving field. For the sake of simplicity, we consider a circularly polarized magnetic driving field $\mathbf{B_1}$ with components $B_{1x} = B_1 \cos(\omega_{MW} t)$ and $B_{1y} = B_1 \sin(\omega_{MW} t)$. The coordinate system (x, y, z) is set up such that the static magnetic field points along the z-axis. Thus, the Larmor precession with frequency $\boldsymbol{\omega}_\mathbf{L}$ takes place in the (x, y) plane. It is convenient to remove the time dependence of $\mathbf{B_1}$ by going into a counterclockwise rotating frame with frequency ω_{MW}. With the offset frequency $\boldsymbol{\Omega}_\mathbf{S} = \boldsymbol{\omega}_\mathbf{L} - \boldsymbol{\omega}_\mathbf{MW}$, Eq. (2.2) takes the following form

$$\begin{aligned}
\frac{\mathrm{d}\mu_x}{\mathrm{d}t} &= -\Omega_S \mu_y\,,\\
\frac{\mathrm{d}\mu_y}{\mathrm{d}t} &= \Omega_S \mu_x - \gamma B_1 \mu_z\,,\\
\frac{\mathrm{d}\mu_z}{\mathrm{d}t} &= \gamma B_1 \mu_y\,.
\end{aligned} \tag{2.10}$$

The presence of the driving magnetic field $\mathbf{B_1}$ induces an additional precession with frequency $\omega_1 = \gamma B_1$. The resulting motion leads to a nutation. In the resonant case $\Omega_S = 0$, the magnetic moment $\boldsymbol{\mu}$ precesses around the x-axis with a frequency ω_1 proportional to the magnitude of the drive field.

Note, that by applying short pulses, one can rotate the magnetization vector by an arbitrary angle. Pulses rotating the spin to the (x, y) plane, are called $\pi/2$-pulses. Accordingly, rotations which invert the magnetization are called π-pulses. These pulses are the essence of pulsed ESR spectroscopy. The frequency ω_1 is also called Rabi frequency. In the quantum mechanical description, the components of the magnetization are replaced by spin operators but the essential dynamics stays the same (see Sec. 2.3.6).

2.3.3 Effective spin and magnetic anisotropy

In order to detect an ion with solid state ESR, it has to possess a partly filled shell. These occur predominantly in the transition groups, i.e. the 3d (iron group), 4d (palladium group), 5d (platinum group), 4f (lanthanide group) and 5f (actinide group). An ion embedded in a solid crystal cannot be approximated by a free ion in vacuum because the ion is surrounded by diamagnetic ions. These so-called ligand ions produce a strong electrostatic field affecting the motion of the electrons. The electrostatic field reflects the local crystal symmetry and the resulting Stark splittings range from 1 THz to several 100 THz. This lifts the degeneracy of the ion's ground state almost entirely. In the case of an ion with an odd number of electrons, Kramer's theorem predicts that at least a twofold degeneracy must remain. These twofold degenerate states are called Kramers doublets. Since the Kramers doublets do not interact with electrostatic fields, it makes sense to introduce the concept of an effective spin 1/2 for describing the doublets. The idea is to keep the form of the Hamiltonian for the Zeeman interaction of the free ion (see Eq. (2.1)) and to introduce a g-tensor accounting for the deviations from the free spin 1/2

$$H = \mu_B \mathbf{S}\mathbf{g}\mathbf{B} \,. \tag{2.11}$$

At this point, it should be noted that similar considerations also hold for the hyperfine and quadrupole interactions, and a hyperfine tensor A and a quadrupol tensor Q have to be introduced, respectively. It is possible to diagonalize the g-tensor. The axes of the diagonalized basis are called (magnetic) principal axes and H takes the following form

$$H = \mu_B \left(g_{xx} B_x \hat{S}_x + g_{yy} B_y \hat{S}_y + g_{zz} B_z \hat{S}_z \right) \,. \tag{2.12}$$

The g-factor for a given orientation of the experiment is given by the direction cosines l, m and n

$$g^2 = l^2 g_{xx}^2 + m^2 g_{yy}^2 + n^2 g_{zz}^2 \,. \tag{2.13}$$

In the special case of cubic symmetry, g remains isotropic. For axial symmetry where $g_{xx} = g_{yy} = g_{\perp}$ and $g_{zz} = g_{\parallel}$, Eq. (2.13) simplifies to

$$g^2 = g_{\parallel}^2 \cos^2\theta + g_{\perp}^2 \sin^2\theta \,. \tag{2.14}$$

Since the principal g values for Er:YSO reveal a low axial anisotropy (e.g. for site 1: $g_x \approx 0$, $g_y \approx 1.5$, $g_z \approx 14.8$ [Sun+08; Gui+06]), Eq. (2.14) is a good approximation to study this spin ensemble. The magnetic anisotropy influences the coupling of the AC probe magnetic field, too. It turns out that the coupling g-factor g_1 for axial symmetry and $B_1 \perp B$ is given by [AB12]

$$g_1 = \frac{g_\perp g_\parallel}{g}. \tag{2.15}$$

Figure 2.2 illustrates the angular dependence of the AC and DC g-factors g_1 and g for axial symmetry.

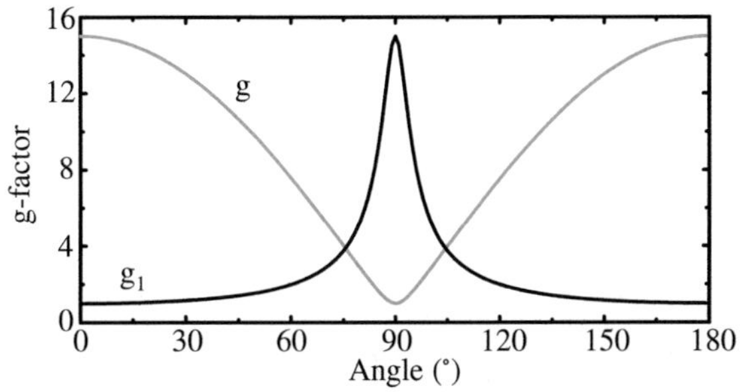

Figure 2.2: Angular dependence of DC and AC g-factors in axial geometry, $g_\perp = 1$, $g_\parallel = 15$.

Note that in contrast to d-group ions, the quenching of the orbital momentum is only a minor effect in the 4f and 5f groups and often leads to a ground state, which carries a large amount of both spin and orbital momentum. This gives rise to large effective g-values and large hyperfine structures. Consequently, those ions are very attractive for applications requiring a large spin tuning rate or operation at zero field and microwave frequencies.

2.3.4 Spin-spin interaction

Spin-spin interaction summarizes all dephasing effect due to the presence of magnetic fields created by the magnetic dipole bath surrounding a single spin. Dipole-dipole interactions broaden the absorption line of the probed spin transitions. An empirical description is the Gaussian approximation for the lineshape

$$f(B) = \frac{1}{\left(2\pi \left\langle B_i^2 \right\rangle\right)^{\frac{1}{2}}} \exp\left(-\frac{\left(B - B_0\right)^2}{2\left\langle B_i \right\rangle^2}\right). \tag{2.16}$$

B_0 is the center of the line, $\left\langle B_i^2 \right\rangle$ is the mean second moment of the line.

On length scales below 0.5 nm, the exchange interaction exceeds the dipole-dipole interaction, which gives rise to ''exchange narrowing'' and results in a Lorentzian line shape of width ΔB

$$f(B) = \frac{1}{\pi}\frac{\Delta B}{(\Delta B)^2 + \left(B - B_0\right)^2}\,. \tag{2.17}$$

However, the later does not apply for the samples investigated in this thesis due to the low concentrations of doped ions. In the context of circuit QED hybrid systems, the linewidth of NV-centers in diamond has been determined to lie between a Gaussian and Lorentzian line shape [San+12; Put+14] and is subject of active research.

In the following, three dephasing mechanisms are presented. It is crucial to understand their origins in a specific material in order to improve the coherence and storage time of a potential quantum memory.

Spectral diffusion

Electronic spin ensembles often feature a rather large linewidth $\Gamma_2^\star$, which can exceed the bandwidth of the excitation pulses used in pulsed ESR spectroscopy. This gives rise to two groups of spins; the A spins which are excited and the B spins which remain in their equilibrium state. The magnetic fields created by the magnetic moments of both A and B spins interact with each other such that spin magnetization gets transferred from the A to the B system where it is undetectable. Also, the equilibrium magnetization can be transferred from the B to the A system. Thus, the observer of the A system notices a relaxation process although non-equilibrium magnetization may be present in parts of the spin ensemble which are not accessible experimentally. This process is called *spectral diffusion*.

Instantaneous diffusion

In experiments with low spin concentration ($< 10^{18}\,\mathrm{cm}^{-3}$), the magnetic interactions between the spins cannot be always neglected. Consider a spin S_1 which gets excited by a microwave pulse. Its precession frequency depends on the external magnetic field. However, if a spin-flip of a neighboring spin S_2 happens, the DC magnetic field slightly changes resulting in a small difference in the precession frequency. Thus, the S_1 spin acquires a different phase compared to its original precession frequency. Multiple spin flips in the environment give rise to a random walk of the S_1 spin frequency through the frequency spectrum. For a randomly distributed spin ensemble of concentration C, the relaxation time caused by *instantaneous diffusion* is given by

$$\frac{1}{T_{\mathrm{ID}}} = C\frac{\pi}{9\sqrt{3}}\frac{\mu_0\mu_B^2 g_1 g_2}{\hbar}\sin^2(\beta/2)\,. \tag{2.18}$$

Here, g_1 and g_2 are the DC g-factors of the excited spins and in most cases $g_1 \approx g_2$ is a good approximation. The angle β quantifies the flipping angle of the spin state due to a microwave pulse. By experimentally varying β, one can separate T_{ID} from other dephasing contributions. For a g-factor of 2 and a concentration of $10^{18}\,\mathrm{cm}^{-3}$, $T_{\mathrm{ID}} \approx 1.2\,\mu\mathrm{s}$ [SE01].

Spin diffusion

Spin diffusion is always present in spin ensembles and is responsible for establishing a common spin temperature. The coupling term $\omega_{ff}(\hat{S}_x^A \hat{S}_x^B + \hat{S}_y^A \hat{S}_y^B)$ mixes A and B spins with a strength given by the flip-flop frequency ω_{ff}. The diffusion rate depends on the distance r of A and B spins and scales with r^{-6}.

2.3.5 Spin-lattice interaction

The goal of this section is to give an order of magnitude estimation of the energy relaxation times of the spins of Kramers ions and determine their functional behavior with temperature. The energy density of the phonon radiation bath is given by

$$\rho_{\mathrm{ph}} \mathrm{d}\omega = \frac{\hbar\omega^3}{2\pi}\left(\frac{2}{v_t^3} + \frac{1}{v_l^3}\right) \frac{\mathrm{d}\omega}{\exp\left(\hbar\omega / k_B T_{\mathrm{ph}}\right) - 1}, \tag{2.19}$$

where v_t and v_l denote the transversal and longitudinal sound velocities, respectively. For this estimation, we will set $v_t = v_l = v$ and ignore any wave velocity dispersion. Phonons propagating through the crystal shake the lattice, which leads to a modulation of the electric field. This gives rise to an orbit-lattice interaction. However, the spin does not react to an electric field and the phonons couple only indirectly to the spins via the spin-orbit interaction. In order to calculate the relaxation rate, it is useful to expand the crystalline electric potential V in powers of the strain $V = V_0 + \epsilon V_1 + \epsilon^2 V_2 + \cdots$. Here, V_0 is the static term, which has no influence on the spin. The matrix element of the first non-vanishing order is

$$\omega_{if} = \frac{2\pi}{\hbar^2}\epsilon^2 \left|\langle i|\, V_1 \,|f\rangle\right|^2 f(\omega). \tag{2.20}$$

Here, $f(\omega)$ is the linewidth function. The states $\langle i|$ and $|f\rangle$ denote the initial and final states of the spin before and after emission of a phonon, respectively. This process is called a *direct process*, where one phonon is emitted or absorbed, respectively. The strain ϵ can be related to the phonon energy density and the Einstein coefficient of induced emission and absorption is given after integrating over the lineshape by $B = \pi |V_1|^2 / \hbar^2 \rho v^2$, where ρ is the crystal density, see [AB12] (chapter 10.4). The temperature dependence of T_1 of a Kramers doublet is given by

$$\frac{1}{T_1} = R_d\,(\hbar\omega)^5 \coth\left(\frac{\hbar\omega}{2k_B T}\right), \tag{2.21}$$

where $R_d = \frac{3}{2\pi\hbar^4 \rho v^5}\frac{1}{\Delta^2}|V_1|^2$ and Δ denotes the crystal field splitting. We now attempt to estimate the T_1 time at a temperature of $T = 20\,\mathrm{mK}$ for an Er^{3+} ion doped into a perfect

Y_2SiO_5 crystal. The density of Y_2SiO_5 is 4.44 g/cm^3 and we assume an average speed of sound of 3×10^3 m/s with an uncertainty of one order of magnitude. For rare earth doped crystals, $|V_1|^2$ is on the order of (300 GHz)^2 [AB12]. The crystal field splitting of the first Er^{3+} Stark level amounts for $\Delta \approx 900$ GHz [Dou+95]. Assuming a Zeeman splitting of 10 GHz, we obtain $T_1 \approx 197$ s. For a Zeeman splitting of 5 GHz, the formula yields $T_1 \approx 105$ min. Note that in our experiments, the crystal lattice is distorted by the comparably large concentration of dopant ions, which alters the transition probabilities and therefore reduces the T_1 time. Nonetheless, energy relaxation times in the order of seconds are to be expected. As a side remark, non-Kramers ions have relaxation times, which are smaller by a factor of 10^{-3}. The reason for the long relaxation times of the Kramers doublets is that in zero magnetic field also the V_1 term vanishes due to the time reversal symmetry. An external magnetic field produces an admixture of excited states, which is on the order of $\hbar\omega/\Delta$ compared to unity for non-Kramers ions. The matrix elements of the spin states of iron group ions are typically larger and the relaxation times shorter [AB12].

The direct process, which was just described, is the dominant relaxation mechanism in the temperature range $k_BT \ll \hbar\omega$ investigated in this thesis. For comparison, we give the full temperature dependence of the longitudinal relaxation for Kramers doublets without further derivation, see [AB12] (chapter 10.4),

$$\begin{aligned} \frac{1}{T_1} &= R_d\,(\hbar\omega)^5 \coth\left(\frac{\hbar\omega}{2k_BT}\right) \text{ (direct process)} \\ &+ R_{\text{or}}\Delta^3 \left[\exp\left(\frac{\Delta}{k_BT}\right) - 1\right]^{-1} \text{ (Orbach process)} \\ &+ R_rT^9 + R_r'\left(\frac{\hbar\omega}{k_B}\right)^2 T^7 \text{ (Raman process)}\,. \end{aligned} \tag{2.22}$$

2.3.6 Quantum mechanical description of the time evolution

Neglecting relaxation for the moment, the time evolution of a quantum system can be described by the Liouville equation

$$\frac{\mathrm{d}\hat{\varrho}}{\mathrm{d}t} = -i\left[\hat{H}(t), \hat{\varrho}\right], \tag{2.23}$$

with the Hamiltonian[2] $\hat{H}(t)$ and the density operator

$$\hat{\varrho} = \sum_i p_i \left|\Psi_i\right\rangle\left\langle\Psi_i\right|. \tag{2.24}$$

If the Hamiltonian is time independent, the time evolution is given by $\hat{\varrho}(t) = \hat{U}\hat{\varrho}(0)\hat{U}^\dagger$ with $\hat{U} = \exp(-i\hat{H}t)$. However, in a typical pulsed ESR experiment, the spin system is

[2] The Hamiltonian is given in units of energy over $\hbar$.

subject to time-dependent perturbations introducing a time dependence into the Hamiltonian. Nonetheless, it is possible to approximate the Hamiltonian by a number of piecewise constant Hamiltonians, such that the approximate time evolution can be written

$$\hat{\varrho}(t) = \hat{U}_n \hat{U}_{n-1} \cdots \hat{U}_1 \hat{\varrho}(0) \hat{U}_1^\dagger \cdots \hat{U}_{n-1}^\dagger \hat{U}_n^\dagger \,. \tag{2.25}$$

The propagators $\hat{U}_i$ represent either pulsed microwave manipulations or periods of free evolution.

Two level dynamics

Here, we consider a spin 1/2 system. The single spin subspace is spanned by the $\hat{S}_x$, $\hat{S}_y$ and $\hat{S}_z$ operators. They are given by the Pauli matrices

$$\hat{S}_x = \begin{pmatrix} 0 & 1/2 \\ 1/2 & 0 \end{pmatrix}, \; \hat{S}_y = \begin{pmatrix} 0 & -i/2 \\ i/2 & 0 \end{pmatrix}, \; \hat{S}_z = \begin{pmatrix} 1/2 & 0 \\ 0 & -1/2 \end{pmatrix} \tag{2.26}$$

The Hilbert space of systems involving multiple spins is given by tensor product of the single spin Hilbert spaces. Figure 2.3 shows the representation of a single spin on the Bloch sphere. The z-component is often referred to as longitudinal magnetization and the projection onto the x-y-plane is called transverse magnetization.

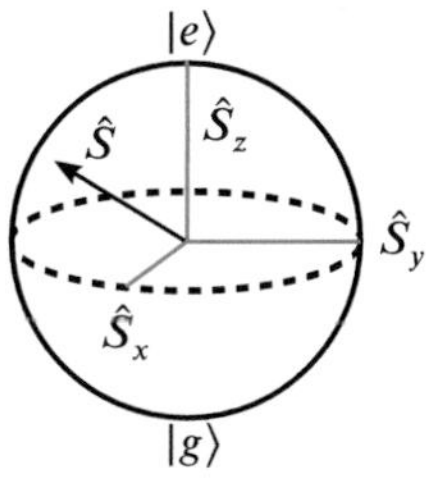

Figure 2.3: Bloch sphere. The kets $|e\rangle$ and $|g\rangle$ denote the excited and ground state of the spin, respectively.

In an external magnetic field, the spin precesses with the Larmor frequency ω_s around the axis of the static magnetic field. For the description of the dynamics close to the resonance, it is convenient to move to a so-called rotating frame by applying the transformation $|\Psi\rangle_{\mathrm{rot}} = \exp(i\omega_{\mathrm{MW}}\hat{S}_z t)\,|\Psi\rangle$. This coordinate system rotates with the applied microwave frequency ω_{MW}. The Hamiltonian of the Zeeman interaction transforms to $\hat{H}_0 = (g\mu_B \Delta B/\hbar)\hat{S}_z = \Omega_s \hat{S}_z$. The Hamiltonian of the oscillatory field applied along the x-axis is given by $\hat{H}_1 = \omega_R \hat{S}_x$, where ω_R is the so-called Rabi frequency.

At zero temperature, the density operator in thermal equilibrium is given by $\hat{\varrho} = -\hat{S}_z$. Then, a $\pi/2$ pulse is applied which rotates the magnetization vector to the $-y$-axis[3]. Thus, the density operator is now given by $\hat{\varrho}_1 = \hat{S}_y$. The time evolution of the density operator follows accordingly

$$\hat{\varrho}_2(t) = e^{-i\Omega_s \hat{S}_z t}\, \hat{\varrho}_1\, e^{i\Omega_s \hat{S}_z t} = \cos(\Omega_s t)\hat{S}_y - \sin(\Omega_s t)\hat{S}_x \,. \tag{2.27}$$

Typically, the precession frequencies of spin ensembles in solids follow a certain distribution, which is often referred to as inhomogeneous broadening. This is taken into account by a normalized distribution $f_{\mathrm{inh}}(\Omega_s)$ and the operators have to be replaced by the ensemble average to infer the net magnetization

$$\overline{\langle \hat{S}_y \rangle} = \frac{1}{2}\int_{-\infty}^{\infty} f_{\mathrm{inh}}(\Omega_s)\cos(\Omega_s t)\, \mathrm{d}\Omega_s \,, \tag{2.28}$$

$$\overline{\langle \hat{S}_x \rangle} = \frac{1}{2}\int_{-\infty}^{\infty} f_{\mathrm{inh}}(\Omega_s)\sin(\Omega_s t)\, \mathrm{d}\Omega_s \,. \tag{2.29}$$

The precessing spins can be compared to oscillating classical magnetic moments representing electromagnetic dipoles. Therefore, the spin ensemble emits electromagnetic radiation proportional to $\overline{\langle \hat{S}_y \rangle}$ and $\overline{\langle \hat{S}_x \rangle}$. Thus, Eqs. (2.28) and (2.29) clearly show that the emitted signal decays over time due to the increased dephasing of the spins until their individual contributions average to zero. This decay is called *free induction decay*.
Up to now, dissipation has been neglected. The presented description can be extended to treat an open quantum system by introducing a Lindblad operator $\hat{L}(\hat{\varrho})$

$$\frac{\mathrm{d}\hat{\varrho}}{\mathrm{d}t} = -i\left[\hat{H}(t), \hat{\varrho}\right] + \hat{L}(\hat{\varrho}) \,, \tag{2.30}$$

The simplest form of $\hat{L}$ for a single spin is given by

$$\hat{L}(\hat{\varrho}) = -\Gamma_1 \hat{S}_z \varrho_e - \Gamma_2\left(\hat{S}_+ \varrho_{eg} + \hat{S}_- \varrho_{ge}\right) \,, \tag{2.31}$$

where $\hat{S}_\pm = (\hat{S}_x \pm i\hat{S}_y)/2$. The indices of ϱ denote the respective entries of the density matrix, where e denotes the excited and g the ground state. The first term in Eq. (2.31) describes the longitudinal relaxation whereas the second term describes the dephasing $\Gamma_2 = \Gamma_1/2 + \Gamma_p$, where Γ_p denotes the pure dephasing rate.

Electron spin echos

Paramagnetic spin ensembles can have a quite large inhomogeneous broadening on the oder of 10 MHz. In the context of quantum memory research, this leads to a fast decay of stored information withing 100 ns due to the free induction decay. The previous analysis has been

[3] The sign of the spin operator for electrons is opposite to the magnetization, e.g. the magnetization component M_x corresponds to $\hat{\varrho} = -\hat{S}_x$.

performed without considering relaxation processes indicating that the information of the initial $\pi/2$-pulse is still in the system. Indeed, if the time evolution could be reversed, all spins would return to their starting position, i.e., being in phase again and emitting a measurable signal. This can be done by applying an additional π-rotation around the x-axis. Such a pulse inverts the phase ϕ of the y-component of the magnetization vector ($\phi \rightarrow -\phi$) and the x-component stays unchanged. The density operator before the second pulse has been derived in Eq. (2.27), and right after the pulse is given by

$$\hat{\varrho}_3(\tau) = -\cos(\Omega_s\tau)\hat{S}_y - \sin(\Omega_s\tau)\hat{S}_x\,, \qquad (2.32)$$

where τ denotes the delay between the $\pi/2$ and the π-pulse. The subsequent evolution reads

$$\hat{\varrho}_4(\tau + t) = -\cos(\Omega_s(t-\tau))\hat{S}_y + \sin(\Omega_s(t-\tau))\hat{S}_x\,. \qquad (2.33)$$

For $t = \tau$, $\hat{\varrho}_4 = -\hat{S}_y$ which means that the system is refocused. This time evolution is now inserted into the equations of the inhomogeneous ensemble yielding $\overline{\langle \hat{S}_y \rangle} = -1/2$ and $\overline{\langle \hat{S}_x \rangle} = 0$. Hence, the system is rephased and emits an echo pulse. This sequence is called *two pulse echo sequence*, see Fig. 2.4 for a schematic representation. Note that we have neglected the case where the inhomogeneous linewidth is larger than the bandwidth of the pulse. In this case, the description becomes more complex in terms of the line shape of the emitted echo. In a realistic system with relaxation, the two pulse echo enables us to measure the coherence time T_2 by varying the time τ between the pulses and recording the emitted echo amplitude. For the analysis of the data, the recorded amplitude vs. 2τ is plotted. The characteristic timescale of the exponential decay in this plot is given by the coherence time T_2. There are a number of different pulse schemes and we refer to Ref. [SE01] for further details.

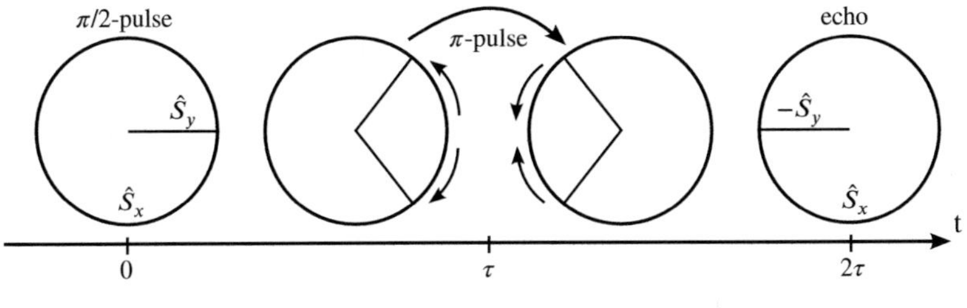

Figure 2.4: From left to right, spin dynamics of the two pulse sequence in the equatorial plane of the Bloch sphere. After the initial $\pi/2$-pulse, the spins defocus. At $t = \tau$, the π-pulse flips the phase of the y-component leading to a rephasing and an echo at $t = 2\tau$.

The second important pulse scheme is the three pulse or *stimulated echo sequence* involving three $\pi/2$-pulses. It can be viewed as a two pulse sequence where the second π-pulse has been separated into two $\pi/2$-pulses with separation time T_W. In a stimulated echo experiment, the first two pulses are separated by a short time τ and the waiting time T_W is varied. The first two $\pi/2$-pulses serve as a polarization generator. This polarization grating is stored for a time

T_W, until it is transferred back to transverse magnetization by a so-called read-pulse with a subsequent echo after a time τ. For the quantitative description, it is again convenient to look at the evolution of a single spin packet. After the second π-pulse, the density operator is given by

$$\hat{\varrho}(\tau) = \cos(\Omega_s \tau)\hat{S}_z \,. \tag{2.34}$$

This is called a polarization grating. Then, after T_W, a final $\pi/2$ pulse yields

$$\hat{\varrho}(\tau + T_W) = -\cos(\Omega_s \tau)\hat{S}_y \,, \tag{2.35}$$

and the following time evolution after time t

$$\begin{aligned} \hat{\varrho}(\tau + T_W + t) &= -\frac{1}{2}\left[\cos(\Omega_s(t-\tau)) + \cos(\Omega_s(t+\tau))\right]\hat{S}_y \\ &+ \left[\sin(\Omega_s(t-\tau)) + \sin(\Omega_s(t+\tau))\right]\hat{S}_x \,. \end{aligned} \tag{2.36}$$

The terms with $\sin(\Omega_s(t+\tau))$ and $\cos(\Omega_s(t+\tau))$ cannot yield an echo response because the time t is only positive valued. However, for $t = \tau$ all terms with $\sin(\Omega_s(t-\tau))$ and $\cos(\Omega_s(t-\tau))$ become time independent and contribute. Note that in the three pulse sequence half of the polarization defocuses such that for the ensemble response one gets $\overline{\langle \hat{S}_y \rangle} = -1/4$. This is half of the two pulse echo signal. The relaxation of the stimulated echo with longer T_W is limited by the so-called phase memory time T_m, which will be further discussed in the experimental part of the thesis.

Electron spin echo envelope modulation (ESEEM)

Electron spin echo envelope modulation (ESEEM) is a pulsed ESR technique, which allows to study the magnetic environment of an electron spin. The electron spin of an ion doped into a crystal interacts with the surrounding nuclear spins by dipole-dipole interaction. Here, we consider a four state system $|S, I\rangle$ of one electron spin $\mathbf{S}$ and a nuclear spin $\mathbf{I}$. The electronic and nuclear precession frequencies are denoted by ω_e and ω_I, respectively. If the nuclear spins are sufficiently far away ($\gg 0.3$ Å), Fermi contact coupling is negligible and the interaction Hamiltonian reads

$$H^{\text{int}} = -\left[\frac{g\mu_B g_n \mu_n \mathbf{S}\cdot\mathbf{I}}{R^3} - \frac{3(g\mu\mathbf{S}\cdot\mathbf{R})(g_n\mu_n\mathbf{I}\cdot\mathbf{R})}{R^5}\right] \tag{2.37}$$

$$\approx \frac{\mu_0}{4\pi}\frac{g\mu_B g_n \mu_n}{r^3}\left[3\cos^2(\theta) - 1\right]\mathbf{S}\cdot\mathbf{I} \,. \tag{2.38}$$

The simplification in the last line is obtained by assuming **S** || **I**, which is justified in a sufficiently strong magnetic field. In this regime, one can evaluate the zz and xz part of the dipole-dipole interaction [Hof89; Sco07]

$$A_{zz} = \frac{\mu_0}{4\pi}\frac{g\mu_B g_n \mu_n}{r^3}\left[3\cos^2(\theta) - 1\right], \tag{2.39}$$

$$A_{xz} = \frac{\mu_0}{4\pi}\frac{g\mu_B g_n \mu_n}{r^3} 3\cos(\theta)\sin(\theta). \tag{2.40}$$

The interaction lifts the twofold degeneracy $|-,\pm\rangle$ and $|+,\pm\rangle$ by the super-hyperfine frequencies

$$\hbar\omega_1 = \sqrt{\frac{A_{xz}^2}{4} + \left(\frac{A_{zz}}{2} - \hbar\omega_I\right)^2}, \tag{2.41}$$

$$\hbar\omega_2 = \sqrt{\frac{A_{xz}^2}{4} + \left(\frac{A_{zz}}{2} + \hbar\omega_I\right)^2}, \tag{2.42}$$

where $\omega_I = g_n\mu_n B/\hbar$ is the nuclear Zeeman frequency.

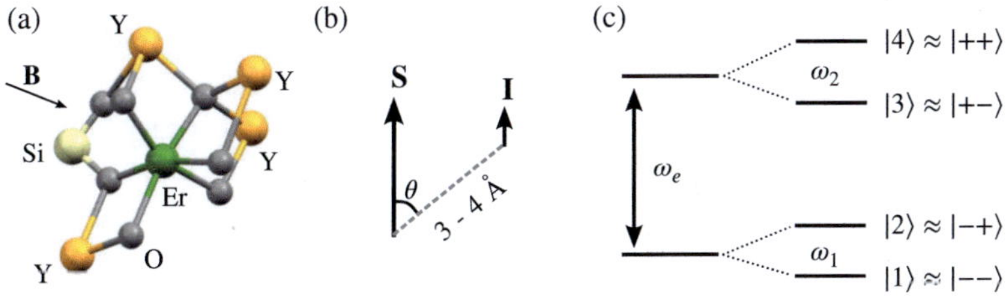

Figure 2.5: (a) Crystal structure of Er:YSO at site 2. (b) Dipole-dipole interaction between an electronic **S** and nuclear **I** spin. (c) Resulting level structure: The electronic levels split by the two super-hyperfine frequencies ω_1 and ω_2.

The new eigenstates deviate slightly from the eigenstates of the uncoupled $|\pm,\pm\rangle$ system. The mixing angles are given by $\phi_1 = A_{xz}/2\omega_1$ and $\phi_2 = A_{xz}/2\omega_2$ yielding

$$|1\rangle = -\sin(\phi_1/2)\,|-+\rangle + \cos(\phi_1/2)\,|--\rangle, \tag{2.43}$$

$$|2\rangle = \cos(\phi_1/2)\,|-+\rangle + \sin(\phi_1/2)\,|--\rangle, \tag{2.44}$$

$$|3\rangle = -\sin(\phi_2/2)\,|++\rangle + \cos(\phi_2/2)\,|+-\rangle, \tag{2.45}$$

$$|4\rangle = \cos(\phi_2/2)\,|++\rangle + \sin(\phi_2/2)\,|+-\rangle, \tag{2.46}$$

where $|1\rangle$ denotes the lowest and $|4\rangle$ the highest energy level.

In the case of Er doped YSO, ^{89}Y features a nuclear spin 1/2 with 100% abundance and $g_n = -0.2748$. The isotopes of Si and O with nuclear spin are negligible due to their low abundance. Here, only even isotopes of Er are considered and the Hamiltonian of an Er-Y pair in a magnetic field reads as follows [Gui+07]

$$H = \mu_B \mathbf{BgS} - \mu_n g_n \mathbf{BI} + H^{\mathrm{int}}, \tag{2.47}$$

where the interaction term is given by the dipole-dipole coupling (see Eq. (2.38)).

For a three pulse echo sequence, the echo intensity for a pair of interacting spins is given by [SE01]

$$I(\tau, T_W) = \frac{1}{2}\left(V_1 + V_2\right), \tag{2.48}$$

where

$$V_1 = 1 - \frac{k}{2}\left[1 - \cos(\omega_2 \tau)\right]\left[1 - \cos(\omega_1(\tau + T_W))\right], \tag{2.49}$$

$$V_2 = 1 - \frac{k}{2}\left[1 - \cos(\omega_1 \tau)\right]\left[1 - \cos(\omega_2(\tau + T_W))\right]. \tag{2.50}$$

The parameter $k = (\omega_I A_{xz})/(\omega_1 \omega_2)$ determines the depth of the modulation and is linked to the transition probabilities between the two allowed and the two partially forbidden transitions [Mim68]. Note, that the two pulse ESEEM signal is given by evaluating Eq. (2.48) with $T_W = 0$. The ESEEM signal can become quite complex if multiple ions are involved and the total intensity for N interacting ions is given by

$$I(\tau, T_W) = \frac{1}{2}\left(\prod_{n=1}^{N} V_{1,n} + \prod_{n=1}^{N} V_{2,n}\right). \tag{2.51}$$

ESEEM spectra are analyzed by removing the baseline (i.e. the exponential decay) and Fourier transforming the data. In this thesis, ESEEM time-domain data is typically multiplied by a Hamming window before a discrete fast Fourier transform is applied. The Hamming window is a window function, which helps to suppress numerical artifacts in the discrete Fourier transform.

2.4 Optical spectroscopy of erbium ions in YSO

Erbium ions doped into Y_2SiO_5 possess an optical transition corresponding to a wavelength of approximately 1.54 µm. The transition lies inside the 4f shell and goes from the $^4I_{15/2}$ to the $^4I_{13/2}$ state as shown in Fig. 2.1(b), Sec. 2.2. The optical transition corresponds to a vacuum wavelength of 1536.48 nm at site 1 and 1538.90 nm at site 2 in zero magnetic field [Böt+06b]. The erbium ions at each site can occupy two different magnetically inequivalent positions, i.e. magnetic classes. Thus, at zero magnetic field, the $^4I_{15/2}$ and $^4I_{13/2}$ states are each twofold degenerate. Applying a magnetic field can lift the degeneracy and a 4-level structure emerges for each class. For each site, 2×4 optical transitions are possible, which are labeled (a^a, b^a, c^a, d^a) for class a and (a^b, b^b, c^b, d^b) for class b. [Sun+08] (see Fig. 2.6).

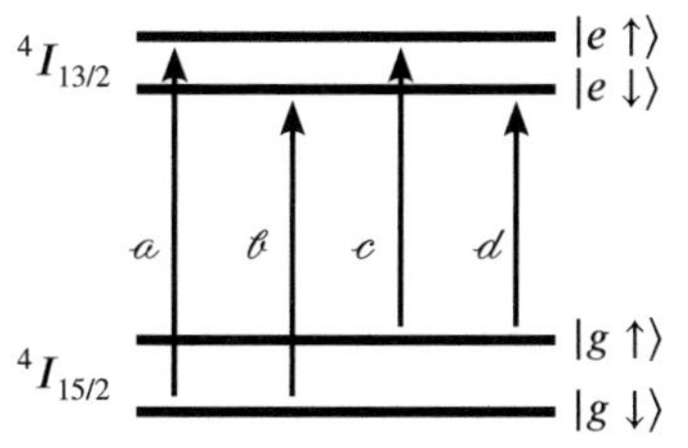

Figure 2.6: Optical transitions for an erbium site and magnetic class. In a magnetic field, the degeneracy of the spin states in the optical ground and excited state is lifted thereby giving rise to four optical transitions.

The g-tensors of the optical ground and excited states differ such that for most orientations of the magnetic field, the transition energies of the optical transitions a and c do not coincide.

2.4.1 CW spectroscopy

In this thesis, the optical spectroscopy is performed in transmission at milli-Kelvin temperatures. At such low temperatures, it is justified to truncate the level structure of Er^{3+} in YSO to the effective 4-level system mentioned above. This is possible because the crystal field splitting is on the order of 1 THz and therefore, $\hbar\omega \gg k_B T$. In the following the optical ground and excited states will be denoted with the letters g and e and the spin state will be indicated by arrows such that in a magnetic field $|g\downarrow\rangle$ denotes the lowest level.
The interaction Hamiltonian of a monochromatic light field (e.g. from a laser) $\mathbf{E}^{(\pm)}(\mathbf{r},t) = \mathbf{E}_0(\mathbf{r},t)\exp(\mp i\omega_0 t)$ with a two level atom is given in the rotating wave approximation by

$$\hat{H}_{\text{rwa}}(t) = -\mathbf{d}\cdot\mathbf{E}_0(t)e^{-i\delta_0 t}\hat{\sigma}_+ - \mathbf{d}^*\cdot\mathbf{E}_0^*(t)e^{i\delta_0 t}\hat{\sigma}_-\,, \tag{2.52}$$

where the envelope of the electric field amplitude $\mathbf{E}_0$ is evaluated at the position of the atom. The $\hat{\sigma}_\pm$ are the raising and lowering operators expressed in the eigenbasis $(|1\rangle, |2\rangle)$. The solution to this problem including dissipation is obtained by solving the Liouville equation, which yields the Bloch equations [GC08]

$$\frac{\mathrm{d}}{\mathrm{d}t}\bar{\varrho}_{22}(t) = -\omega_{21}\bar{\varrho}_{22}(t) + i\left[\Omega(t)\bar{\varrho}_{12}(t) - \Omega(t)^*\bar{\varrho}_{21}(t)\right], \tag{2.53}$$

$$\frac{\mathrm{d}}{\mathrm{d}t}\bar{\varrho}_{11}(t) = \omega_{21}\bar{\varrho}_{22}(t) - i\left[\Omega(t)\bar{\varrho}_{12}(t) - \Omega(t)^*\bar{\varrho}_{21}(t)\right], \tag{2.54}$$

$$\frac{\mathrm{d}}{\mathrm{d}t}\bar{\varrho}_{21}(t) = \left(i\delta - \Gamma_{21}\right)\bar{\varrho}_{21}(t) + i\Omega(t)\left(\bar{\varrho}_{11}(t) - \bar{\varrho}_{22}(t)\right). \tag{2.55}$$

Here, $\omega_{21} = 1/T_1$ is the decay rate, $\Gamma_{21} = 1/T_2$ denotes the dephasing rate and ϱ is the density matrix. The Rabi frequency $\Omega(t) = \mathbf{d} \cdot \mathbf{E}_0(t)/\hbar$ is determined by the field amplitude of the incoming monochromatic light. The optical homogeneous linewidth measured for Er:YSO is 73 Hz at 1.5 K and 7 T [Böt+06b; Mac+97]. To our knowledge, this is the narrowest linewidth measured for an optical transition in any solid [Sun+08; Boe+03].

However, the linewidth measured in experiments is on the order of 100 MHz because the inhomogeneous linewidth dominates the absorption profile. Local variations of the crystal field lead to a distribution of transition frequencies enhancing the dephasing.

2.4.2 Time resolved spectroscopy

In order to determine the coherence properties of the optical transition, one has to study the time evolution of the ensemble. This is done by applying certain resonant optical pulse sequences to the ensemble and measure the response. A π-pulse rotates the Bloch vector by 180°, a $\pi/2$-pulse by 90°. These pulses have to be very short, i.e. much shorter than any decoherence mechanism. Therefore, in experiments, one tries to minimize the pulse length and to maximize the field amplitude of the laser light. One distinguishes three basic pulse schemes, free induction decay, 2 photon echo (2PE) and 3 photon echo (3PE). Since the quantum mechanical description of pulsed microwave spectroscopy has already been discussed in Sec. 2.3.6, it is only briefly summarized for the optical spectroscopy. In optics, one detects intensity rather than amplitude and phase.

Free induction decay

An initial $\pi/2$-pulse brings the Bloch vector of all dipoles to the equatorial plane, where they precess with their transition frequencies. First, all dipoles are in phase but due to the inhomogeneous broadening, they dephase quickly as they evolve with time. Thus, the photo detector will detect an exponentially decaying radiation intensity from the sample.

$$I(t) \propto \exp\left(-2t/T_2^*\right) . \tag{2.56}$$

Since the inhomogeneous broadening is on the order of 100 MHz, this decay is very fast and cannot be observed with our experimental setup [Rie13]. The characteristic timescale of this decay is denoted by T_2^*.

Two pulse echo

The goal of this sequence is to refocus the dipoles by inversing the time evolution. The first $\pi/2$ pulse brings the dipoles to the equatorial plane where they dephase. After a time τ, a π-pulse inverts the magnetization effectively reversing the time evolution, such that after another delay time τ all dipoles are in phase again. This leads to a coherent emission of light. By varying the delay time τ, an exponential decay of the detected intensity is observed

$$I(\tau) \propto \exp\left(-4\tau/T_2\right) . \tag{2.57}$$

Three pulse echo

This sequence allows us to measure the longitudinal relaxation time. First, two $\pi/2$-pulses are applied, which are separated by a short time $\tau \ll T_2$. The dipoles are now slightly dephased and point to the north pole of the Bloch sphere. Next, one waits for a time T_W, where the dipoles are subject to longitudinal relaxation. A final $\pi/2$-pulse brings the ensemble back to the equatorial plane, where the dipoles rephase after a time τ and emit a pulse of light. By varying the time T_W, one can determine the T_1 relaxation

$$I(T_W) \propto \exp\left(-2T_W/T_1\right) . \tag{2.58}$$

Note that this only holds true when effects such as spectral diffusion are negligible.

Spectral diffusion

Spectral diffusion additionally perturbs the optical transition (see also Sec. 2.3.4). It results from the modulation of the transition frequencies of an ion due to the interaction with its dynamic environment. The random variation of the transition frequency of the ion can be viewed as a random walk in frequency or a diffusion through the optical spectrum. In other words, an initial sub-ensemble of dipoles with linewidth Γ will acquire a larger linewidth over time. The dynamic environment can be spins in the vicinity of the ion, structural rearrangement of the crystal lattice or moving charge defects [Böt+06b]. In the case of RE ions at low temperatures, spin-spin interactions represent the dominant mechanism. Here, the implications of this process on the shape of the 3PE spectrum are discussed. In the presence of spectral diffusion, the decay curve of the 3PE experiment deviates from the usual exponential form

$$I(\tau, T_W) = I_0 \exp\left(-2T_W/T_1\right) \exp\left(-4\tau\pi\Gamma_{\text{eff}}(\tau, T_W)\right) . \tag{2.59}$$

Here, τ is the time beween the first two $\pi/2$-pulses and T_W is the waiting time after that until the final $\pi/2$-pulse. The first exponential decay in the equation is the usual T_1 decay from Eq. (2.58) and $\Gamma_{\text{eff}}(\tau, T_W)$ is given by

$$\Gamma_{\text{eff}}(\tau, T_W) = \Gamma_0 + \frac{1}{2}\Gamma_{\text{SD}}\left[R\tau + \left\{1 - \exp\left(-RT_W\right)\right\}\right] . \tag{2.60}$$

The first summand Γ_0 denotes the linewidth in the absence of spectral diffusion, Γ_{SD} is the FWHM linewidth broadening due to dipole-dipole interactions and R is the rate of the diffusion process, which is given by the spin-flip transition rates of the perturbing spins [Böt+06b].

The FWHM linewidth due to dipole-dipole interactions is given by [Böt+06b]

$$\Gamma_{\max} = \frac{\pi}{9\sqrt{3}} \frac{\mu_0 \left|g_g - g_e\right| g_{\text{env}} \mu_B^2 n_{\text{env}}}{h} , \tag{2.61}$$

assuming an isotropic g-tensor of the spins and the environment. The spectral diffusion rate can be calculated as follows

$$\Gamma_{SD}(B,T) = \Gamma_{max} \mathrm{sech}^2 \left(\frac{g_{env} \mu_B B}{2 k_B T} \right) . \tag{2.62}$$

Here, B denotes the static magnetic field, T is the temperature and g_{env} is g-factor of the magnetic environment.

3 Hybrid quantum circuits

This chapter provides an introduction into hybrid quantum circuits based on superconducting resonators and qubits coupled to paramagnetic spin ensembles. First, superconducting transmission lines, microwave resonators and qubits are introduced. Then, the coupling mechanism between those components and paramagnetic spin ensembles is presented.

3.1 Superconducting quantum circuits

Superconducting (SC) quantum circuits are macroscopic electrical circuits fabricated with lithographic techniques that operate in the quantum regime. The building blocks consist of electromagnetic LC harmonic oscillators, working at microwave frequencies, and non-linear resonators, forming the qubits. In this architecture, the harmonic LC resonators serve as storage, coupling or readout units for the qubits. Such an approach is extremely versatile because it allows the engineering of specific quantum systems. Those circuits may serve as *on-chip* quantum simulators to compute complex quantum systems [GAN14]. The field of this so-called SC circuit quantum electrodynamics (QED) is rapidly evolving. The ability to control and create almost arbitrary quantum states [Hof+09] has led to the realization of prototype quantum computers [Luc+12; Bar+14]. A short review of the state of the art is given in Ref. [DS13]. The goal of this thesis is to explore the coupling of SC circuits to rare earth ions, which can serve as long coherent memories or reversible quantum converters for SC qubits.

3.1.1 Superconductivity

In 1911, H. Kammerlingh Onnes noticed, that for certain metals and below a (material dependent) critical temperature, the electrical resistance vanishes [Onn11]. This interesting property is accompanied by perfect diamagnetism discovered by Meißner and Ochsenfeld in 1933 [MO33]. These findings show that superconductivity is a new thermodynamic state described by a complex order parameter Ψ [GL50]. The microscopic origin of type I superconductivity is explained by the BCS theory [BCS57]. It turns out that the order parameter Ψ describes a macroscopic wave function $\Psi = \sqrt{n_S}\exp(i\varphi)$, where n_S is the density of the superconducting condensate (the Cooper pairs) and φ is the phase. In contrast, the so-called high-T_c or type II superconductivity is still an active field of research.

For the purpose of this thesis, it is sufficient to remember that superconducting circuits are low loss, electrical circuits that are described by a macroscopic wave function, and therefore, they can exhibit quantum phenomena on a macroscopic level.

3.1.2 Qubits

The research on superconducting quantum information circuits started with the realization of the charge qubit [NPT99; Bou+98], which consists of a superconducting island connected by a so-called Josephson junction and a capacitor to the rest of the circuitry. The Josephson junction is a crucial non-linear circuit element and it will by described in detail in the next paragraph. The charge qubit has been improved over the years to the so-called transmon qubit [Koc+07; Sch+07], which is widely used in todays scientific research. Other types of qubits are the phase qubit [MDC85; Yu+02] and the flux qubit with one Josephson junction [Fri+00]. Modern versions of the flux qubit employ three Josephson junctions [Moo+99] and the fluxonium qubit is made of a large number of junctions [Man+09]. Among those qubits, the three junction flux qubit is the most interesting candidate for directly interfacing qubits with spin ensembles in solids [Zhu+11]. In the following, this qubit will be discussed in greater detail.

A quantum harmonic oscillator has an even energy level spacing with a separation given by $\hbar\omega$, where $\omega/2\pi$ denotes the frequency. By applying a (classical) microwave pulse with frequency $\omega/2\pi$, the oscillator gets excited. However, due to the even level spacing, it is not possible to address a certain level individually. Instead, the excitation gets distributed over all levels following a Poisson distribution. In order to form a qubit, a non-linearity is required, which makes those levels distinguishable in energy. This allows to truncate this anharmonic oscillator to an effective two level system. The most convenient non-dissipative non-linearity in SC circuits is a Josephson junction [Jos62]. It consists of two superconducting electrodes separated by a thin insulating barrier. Inside this barrier, the macroscopic wavefunction decays exponentially. Since the system is described by a single wave function and assuming the same electrode material on both sides, the wave functions on either side of the barrier only differ by a phase φ. This leads to the Josephson equations

$$I(\varphi) = I_c \sin(\varphi), \tag{3.1}$$

$$\frac{\partial\varphi}{\partial t} = \frac{2\pi}{\Phi_0} U, \tag{3.2}$$

with the flux quantum $\Phi_0 = h/2e$ and the critical current I_c. By inserting the Josephson equations into the definition of the response of an inductor $U = -L\mathrm{d}I/\mathrm{d}t$, one obtains th non-linear inductance of the Josephson junction

$$L_J(\varphi) = \frac{\Phi_0}{2\pi I_c \cos(\varphi)} \tag{3.3}$$

The energy associated with a Josephson junction is $E = \int I U \mathrm{d}t$ resulting in

$$E = E_J(1 - \cos\varphi), \tag{3.4}$$

with $E_J = \hbar I_c/2e$ after inserting the Josephson equations. Figure 3.1(a) shows a micrograph of a flux qubit fabricated by O. Astafiev similar to the one analyzed in Ref. [Abd+08].

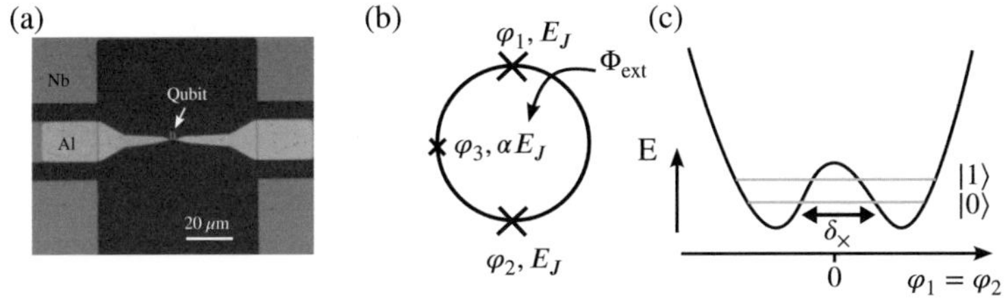

Figure 3.1: (a) Micrograph of a flux qubit (small circular structure) embedded in a coplanar microwave resonator (central line). (b) Schematics of a three junction flux qubit. (c) Effective potential of the flux qubit with its two computational states (gray lines).

In the following, we will briefly derive the basic properties of the flux qubit. A detailed description is for instance found in Ref. [Jer13]. The qubit is schematically presented in Fig. 3.1(b) and consists of a superconducting loop interrupted by three Josephson junctions, two junctions are identical and one has an α-times smaller Josephson Energy E_J. The SC wave function has to be single valued along a closed loop. This means that by going around the loop, the phase is only allowed to acquire integer multiples of 2π. In other words, the currents flowing through the loop are quantized, as well as the magnetic flux they create. This gives rise to the so-called flux quantization condition. Taking the phase changes at the Josephson junctions into account, the following flux quantization condition has to be fulfilled

$$\sum_{i=1,2,3} \varphi_i + 2\pi \frac{\Phi_{\text{ext}}}{\Phi_0} = 2\pi N\,. \tag{3.5}$$

Here, Φ_{ext} denotes the externally applied magnetic flux and N is the number of flux quanta in the loop. Equation (3.5) allows to eliminate one degree of freedom and the potential energy is given by [Moo+99]

$$\frac{E}{E_J} = 2 + \alpha - \cos(\varphi_1) - \cos(\varphi_2) - \alpha\cos(2\pi\Phi_{\text{ext}}/\Phi_0 - \varphi_1 - \varphi_2)\,. \tag{3.6}$$

For $0.5 < \alpha < 1.0$, a flux qubit is formed. The corresponding (2π periodic) potential shows two minima along the axis defined by ($\varphi_1 = \varphi_2$) for $\Phi_{\text{ext}}/\Phi_0 = 0.5$, see Fig. 3.1(c). These two (classically) degenerate groundstates are characterized by persistent supercurrents flowing in opposite directions in the loop of magnitude

$$I_p = I_c\sqrt{1 - \frac{1}{4\alpha^2}}\,. \tag{3.7}$$

These persistent currents can be quite large, creating a macroscopic magnetic moment, which makes flux qubits particularly interesting for hybrid quantum systems with natural spins. Due to the quantum mechanical tunnel coupling $\delta_\times$ between both groundstates, the degeneracy is lifted, which forms the two computational states of the flux qubit $|0\rangle$ and $|1\rangle$.

3.1.3 Transmission lines and microwave resonators

Transmission lines (TL) are coplanar waveguides (CPW) used to guide the microwave signals on SC chips. They consist of a central conductor, which is separated by gaps on both sides from the electrical ground, see Fig. 3.2(a). One can regard these TLs as two dimensional versions of the more common coaxial cables used in microwave communication. In our circuits, they are usually impedance matched to 50 Ω and Fig. 3.2(a) presents the equivalent circuit [Poz05]. The impedance of the lossless CPW is given by $Z = \sqrt{L/C}$, which is a valid approximation because we are mostly concerned with SC transmission lines.

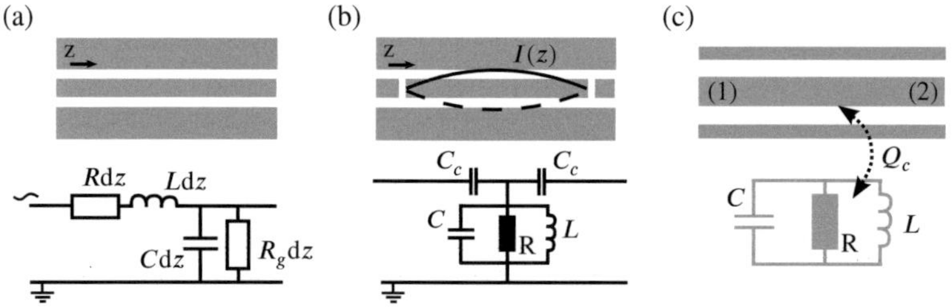

Figure 3.2: (a) Coplanar transmission line and corresponding equivalent circuit. (b) Transmission line resonator. The solid and dashed lines sketch the current distribution of the first harmonic. The lumped element circuit model of the resonator is shown below. (c) Lumped element resonator coupled to a transmission line (notch type configuration).

By interrupting the central conductor at two locations by a small gap, a so-called CPW resonator is formed, see Fig. 3.2(b). Its resonance frequency and higher harmonics are determined by the boundary conditions set by the two gaps giving rise to standing waves.

In addition to these distributed element resonators, lumped element (LE) resonators will be employed in the investigations presented in this theses. They consist of spacially separated inductances and capacitances and they are typically smaller in size compared to CPW resonators. Due to their small size, it is convenient to arrange multiple LE resonators on a single chip with different resonance frequencies [Wue+11]. Figure 3.2(c) shows a typical coupling geometry where an LE resonator is coupled to a transmission line.

It is convenient to describe elements of microwave networks in terms of a scattering matrix S. The notch type resonator presented in Fig. 3.2(c) represents a two port network indicated by the numbers (1) and (2) in the figure. Let $V_i^{\pm}$ denote the amplitude of the voltage wave at port i and + and − the incident and reflected wave at that port, respectively. The system is completely described by [Poz05]

$$\begin{pmatrix} V_1^- \\ V_2^- \end{pmatrix} = \begin{pmatrix} S_{11} & S_{12} \\ S_{21} & S_{22} \end{pmatrix} \begin{pmatrix} V_1^+ \\ V_2^+ \end{pmatrix} . \tag{3.8}$$

Typically, we probe the transmission coefficient S_{21} in our experiments. The complex S_{21} transmission through the system displayed in Fig. 3.2(c) is given by [Kha+12; Gao08]

$$S_{21}(f) = \underbrace{ae^{i\alpha}e^{-2\pi i f \tau}}_{\text{environment}} \underbrace{\left[1 - \frac{\left(Q_l/|Q_c|\right)\, e^{i\phi}}{1 + 2iQ_l\left(f/f_r - 1\right)}\right]}_{\text{ideal resonator}} . \tag{3.9}$$

Here, Q_l and Q_c describe the loaded and coupling quality factors, respectively, f_r is the resonance frequency and ϕ denotes a possible impedance mismatch. The quantities labeled "environment" take into account the cable delay τ, additional attenuation or amplification a present in the measurement setup as well as an additional phase α. The internal quality factor is given by $Q_i^{-1} = Q_l^{-1} - \mathrm{Re}\left\{Q_c^{-1}\right\}$. Since the internal quality factor describes intrinsic losses, it can also be used to detect the presence of weakly coupled spins by monitoring the change of Q_i as a function of the magnetic field. In this thesis a robust and efficient algorithm to fit Eq. 3.9 has been developed and published in Ref. [Pro+14b].

3.2 Coupling spin ensembles with microwave circuits

3.2.1 General considerations

Resonators

The coupling strength of a single spin to a cavity is determined by the field strength of the vacuum fluctuations. The energy of the ground state $|0\rangle$ of the cavity is given by

$$\langle 0|\int \frac{1}{\mu_0}|B_c|^2\,\mathrm{d}^3r\;|0\rangle = \frac{\hbar\omega}{2} . \tag{3.10}$$

Here, $\hat{B}_c = iB_0\left[f(r)\hat{a} - f^*(r)\hat{a}^\dagger\right]$ denotes the magnetic field operator with the normalization factor B_0 and Eq. (3.10) simplifies to [HR13]

$$\langle 0|\;\frac{1}{\mu_0}B_0^2\,V_m\;|0\rangle = \frac{\hbar\omega}{2} , \tag{3.11}$$

where V_m denotes the effective mode volume given by $V_m = \int\; |f(r)|^2\mathrm{d}^3r$. The normalization factor B_0 represents the root mean square (r.m.s.) magnetic field amplitude of the vacuum fluctuations, which will be denoted by B_{vac} from now on and can be calculated from Eq. (3.11) [HR13]

$$B_{\mathrm{vac}} = \sqrt{\frac{\mu_0\hbar\omega}{2V_m}} . \tag{3.12}$$

The coupling of a single spin 1/2 to the cavity is given by the product of its magnetic moment and the vacuum field

$$\upsilon_1 = \frac{1}{2}\mu_B g B_{\mathrm{vac}}/\hbar\,, \tag{3.13}$$

where μ_B is the Bohr magneton and g is the effective g-factor. For typical dimensions of SC resonators, a single spin coupling on the order of 10 Hz is obtained. This is by far smaller than the typical dissipation rates of SC circuits, which are on the order of 100 kHz.

For lumped element resonators, the coupling strength can be estimated using the inductance and the magnetic field dependence on the current. The average current through the inductance L per one oscillation period is $I = \sqrt{E/4L}$, where E denotes the electromagnetic energy. In order to estimate the fluctuation current, we set $E = \hbar\omega/2$ with the resonance frequency $\omega/2\pi$ of the circuit. Using this current, one can deduce the magnetic field and insert it into Eq. (3.13) yielding the coupling strength.

Flux Qubits

Similar considerations hold for the flux qubit. The vacuum fluctuation current is determined by the persistent current in the qubit loop. Typical values are in the region of $I_p = 150$ nA. The absolute value of the magnetic field created by the persistent current is given by $B = \mu_0 I_p/2R$ to first order. Typical coupling strengths between single spins and flux qubits are about a factor of 10^3 larger compared to lumped element resonators. Since the persistent current is controlled by the design of the Josephson junctions, it should be possible increase it up to 500 nA, while simultaneously decreasing the geometric dimensions to boost the coupling strength. Assuming a diameter of the wire forming the loop of 20 nm and a single spin situated at 20 nm distance, the coupling would already amount for ~ 70 kHz. In order to enter the strong coupling regime, a coherence time of 20 µs would be sufficient, which is already achieved today [Ste+14].

However, typical modern SC circuits operate in a regime where single spins cannot be accessed coherently. The following sections provide a solution to this problem by employing a whole ensemble of of identical spins, which boosts the coupling strength by the square root of their number.

3.2.2 The Tavis-Cummings model

The Tavis-Cummings model [TC68] describes the interaction between N identical two-level systems (spins) and one cavity mode. In the rotating wave approximation, it is given by

$$\hat{H}_{\mathrm{TC}} = \underbrace{\hbar\omega_c \hat{a}^\dagger \hat{a}}_{\text{cavity}} + \underbrace{\sum_{k=1}^{N} \frac{1}{2}\hbar\omega_s \hat{\sigma}_k^z}_{\text{spins}} + \underbrace{\sum_{k=1}^{N} \hbar\upsilon_1 \left(\hat{a}\hat{\sigma}_k^+ + \hat{a}^\dagger \hat{\sigma}_k^-\right)}_{\text{interaction}}\,. \tag{3.14}$$

Here, ω_c denotes the cavity frequency, ω_s is the spin transition frequency and υ_1 is the single spin-cavity coupling strength. The Hamiltonian commutes with the total number of excitations present in the system $\hat{n} = \hat{a}^\dagger\hat{a} + \sum_{k=1}^{N} |\uparrow\rangle\langle\uparrow|$, which allows diagonalization in the eigenspaces of constant excitation [Bre09]. For $n = 1$, the basis states $|\text{cavity; spins}\rangle$ are $|1; \downarrow \cdots \downarrow\rangle$ and all N states of the form $|0; \downarrow \cdots \uparrow_i \cdots \downarrow\rangle$. Within this manifold, diagonalization of the Hamiltonian yields the following dressed states

$$|+\rangle = \cos(\vartheta)\,|1; \downarrow \cdots \downarrow\rangle + \sin(\vartheta) \sum_{k=1}^{N} \frac{1}{\sqrt{N}} |0; \downarrow \cdots \uparrow_k \cdots \downarrow\rangle\,, \tag{3.15}$$

$$|-\rangle = -\sin(\vartheta)\,|1; \downarrow \cdots \downarrow\rangle + \cos(\vartheta) \sum_{k=1}^{N} \frac{1}{\sqrt{N}} |0; \downarrow \cdots \uparrow_k \cdots \downarrow\rangle\,, \tag{3.16}$$

and the eigenenergies

$$E_\pm = \hbar\frac{\omega_c + \omega_s}{2} \pm \hbar\sqrt{\frac{(\omega_c - \omega_s)^2}{4} + \upsilon_N^2}\,. \tag{3.17}$$

The mixing angle is given by $\tan(2\vartheta) = (2\upsilon_1\sqrt{N})/(\omega_c - \omega_s)$ and the collective coupling strength $\upsilon_N = \upsilon_1\sqrt{N}$ has been introduced [Bre09]. The rest of the remaining $N-1$ eigenstates have eigenenergies of $\hbar\omega_s$ and are antisymmetric combinations of the $|0; \downarrow \cdots \uparrow_i \cdots \downarrow\rangle$ states that do not couple directly to the cavity which is why they are called *dark states*. Equation (3.17) shows that a collective coupling strength emerges, which increases the resonator - spin coupling by the square root of the number of spins. Even if the single spin coupling can be very small ($\sim 10\,\text{Hz}$), a large number of spins increases the coupling to the MHz regime allowing strong coupling. An excitation transferred to the spin system manifests itself as a coherent symmetric superposition of all $|0; \downarrow \cdots \uparrow_i \cdots \downarrow\rangle$ states. For the second and higher order excitation manifolds the same considerations hold as long as the number of excitations is much smaller than the number of spins. In this low excitation regime, the situation presents itself as two coupled harmonic oscillators. This is in strong contrast to the case of a single spin coupled to a cavity, where the level structure is non-linear [Fin+08].

Of course, the Tavis-Cummings model is a simplified model. For instance, the spin transition frequencies are not identical in an actual doped crystal but rather follow a distribution. The following sections address the case of an inhomogeneously broadened spin ensemble coupled to a microwave resonator in the weak and strong coupling regime.

3.2.3 Weak Coupling

In a weakly coupled spin ensemble, the coupling strength υ_N of the microwave resonator to the spin ensemble is smaller than the linewidths[1] of both resonator and spin ensemble $\upsilon_N \ll \kappa, \Gamma_2^\star$. The magnetic susceptibility of the spin system modifies the inductance of the

[1] Note that all linewidths are HWHM linewidth if not explicitly mentioned otherwise.

microwave resonator to $L' = L(1+\chi\zeta)$, where, ζ is a filling factor accounting for the specific mode distribution of the magnetic mode of the resonator. In the limit of $\chi\zeta \ll 1$, one can approximate the expression for the modified resonance frequency of the microwave resonator by a Taylor expansion with respect to $\chi\zeta$

$$\omega' = \frac{1}{\sqrt{L'C}} \approx \omega_0 - \frac{1}{2}\omega_0\chi\zeta\,, \tag{3.18}$$

where $\omega_0 = 1/\sqrt{LC}$ is the unperturbed resonance frequency. The susceptibility of a paramagnetic spin ensemble is obtained by solving the Bloch equations in the steady state where the change of the magnetization $\frac{\mathrm{d}}{\mathrm{d}t}M_z$ is zero. In the limit of low probing power, the real and imaginary parts of the susceptibility are given by [AB12]

$$\chi' = -\chi_0\frac{\omega_s\Delta}{\Delta^2+\Gamma_2^{\star 2}}\,, \tag{3.19}$$

$$\chi'' = \chi_0\frac{\omega_s\Gamma_2^{\star}}{\Delta^2+\Gamma_2^{\star 2}} \tag{3.20}$$

Here, χ_0 denotes the DC susceptibility and $\Delta = \omega_0 - \omega_s$ is the detuning. Let us introduce the coupling constant $\upsilon = \omega_0\sqrt{\chi_0\zeta/2}$, which quantifies the energy exchange between the resonator and the spin system. This is justified by the following consideration: In the limit of a detuning larger than the linewidth $\Gamma_2^{\star}$, no energy is exchanged between the two systems. In this so-called dispersive limit, the dispersive frequency shift is given by υ^2/Δ which is compared to Eq. (3.18) combined with Eq. (3.19)

$$\frac{\upsilon^2}{\Delta} = \frac{\chi_0\zeta\,\omega_0\omega_s}{2}\frac{\Delta}{\Delta^2+\Gamma_2^{\star 2}} \approx \frac{\chi_0\zeta\,\omega_0^2}{2\,\Delta}\,. \tag{3.21}$$

Inserting the DC susceptibility $\chi_0 = M\mu_0/B$ yields $\upsilon_N = g\mu_B\sqrt{\mu_0\omega_0 n\zeta/4\hbar}$, where n denotes the density of spins. Thus, the modified resonance frequency and linewidth of the microwave cavity is given by [Bus+11]

$$\omega_0' = \omega_0 + \frac{\upsilon^2\Delta}{\Delta^2+\Gamma_2^{\star 2}}\,, \tag{3.22}$$

$$\kappa_0' = \kappa_0 + \frac{\upsilon^2\Gamma_2^{\star}}{\Delta^2+\Gamma_2^{\star 2}}\,. \tag{3.23}$$

Resonance frequencies and linewidths are determined for instance by using Eq. (3.9), where $Q_l = \omega_0/2\kappa$. Note that Eqns. (3.22) and (3.23) are only valid in the weak coupling regime where just a single resonance is present.

3.2.4 Strong coupling and inhomogeneous broadening

In order to describe an actual experiment, the Tavis Cummings Hamiltonian is not sufficient. The cavity of the hybrid system is connected to the outside world and probed either in transmission or reflection. Therefore, it is an open quantum system. In input-output theory [GC08] (chapter 14.3), this is modeled by bath operators $\hat{c}_i$, $\hat{c}_r$ and $\hat{c}_t$ describing the incident, reflected and transmitted field. Figure 3.3 schematically illustrates the situation. The signal detected by a vector network analyzer can be the complex transmission through the system $S_{21}(\omega) = \langle \hat{c}_t \rangle / \langle \hat{c}_i \rangle$, which is given by the ratio of the expectation values of the transmitted and incident bath operators. A measurement in reflection is also possible, which is described by $S_{11}(\omega) = \langle \hat{c}_r \rangle / \langle \hat{c}_i \rangle$. Here, the calculation of $S_{21}(\omega)$ will be sketched, which is similar to the $S_{11}(\omega)$ case.

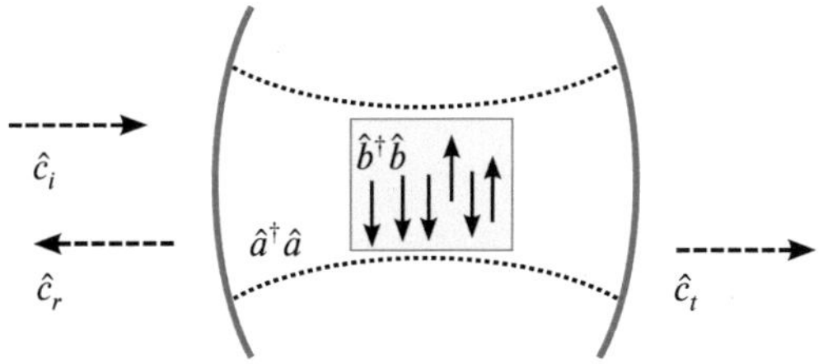

Figure 3.3: A cavity with a spin ensemble coupled to the environment.

The cavity decay rate is given by 2κ, where κ is the HWHM linewidth. The resonance frequency of the cavity is $\omega_c/2\pi$ The decay rate of spins into modes other than the cavity mode is given by 2γ (HWHM homogeneous spin linewidth). The spin ensemble contains N spins with a certain distribution of transition frequencies $\omega_k/2\pi$ around a mean frequency $\omega_S/2\pi$. The system is considered in the few photon limit, where the spin operators $\hat{s}_k$ can be replaced by bosonic operators $\hat{b}_k$ (Holstein-Primakoff approximation [HP40]). The Heisenberg equations of motion written in the rotating frame of the incident field at frequency $\omega/2\pi$ are given by [Din+11]

$$\begin{aligned}
\dot{\hat{a}} &= -\left[\kappa + i\left(\omega_c - \omega\right)\right]\hat{a} - \sqrt{\kappa}\,\hat{c}_i + \sum_k v_k \hat{b}_k + \hat{f}_a(t)\,, \\
\dot{\hat{b}}_k &= -\left[\gamma + i\left(\omega_k - \omega\right)\right]\hat{s}_k - v_k\hat{a} + \hat{f}_k(t)\,, \\
\hat{c}_r &= \hat{c}_i + \sqrt{\kappa}\,\hat{a}\,, \\
\hat{c}_t &= \sqrt{\kappa}\,\hat{a}\,.
\end{aligned} \tag{3.24}$$

Here, v_k denotes the coupling strength of a spin to the cavity and $\hat{f}_a(t)$ and $\hat{f}_k(t)$ are noise operators ensuring the preservation of the commutation relations. In the limit of large N, the spin distribution can be approximated by a continuous distribution and the complex transmission is given by [Din+11]

$$S_{21}(\omega) = \frac{\langle \hat{c}_t \rangle}{\langle \hat{c}_i \rangle} = \frac{\kappa / i}{\omega - \omega_c + i\kappa - W(\omega)}, \tag{3.25}$$

with

$$W(\omega) = v_N^2 \int_{-\infty}^{\infty} \frac{\rho(\omega')}{\omega - \omega' + i\gamma} \mathrm{d}\omega', \tag{3.26}$$

where ρ is the normalized distribution of the spin frequencies (inhomogeneous broadening) and $v_N = \sqrt{N} v_1$ is the collective coupling strength. Typically, the homogeneous linewidth γ is small, especially in the case of rare earth spin ensembles, and can be neglected in most applications due to the large inhomogeneous broadening. In the absence of inhomogeneous broadening $\rho(\omega) = \delta(\omega - \omega_S)$, Eq. (3.26) is easily computed, but more realistic cases may be Lorentzian and Gaussian spin distributions, respectively [Din+11; Afz+13]:

$$\begin{aligned} W_\delta(\omega) &= \frac{v_N^2}{\omega - \omega_s + i\gamma} \\ W_{\mathrm{Lorentz}}(\omega) &= \frac{v_N^2}{\omega - \omega_s + i\gamma + i\Gamma_2^\star} \\ W_{\mathrm{Gauss}}(\omega) &= -i\frac{\sqrt{\ln 2}\, v_N^2}{2\Gamma_2^\star} \sqrt{\pi}\, \exp\left[-\left(\frac{(\omega - \omega_s) + i\gamma}{2\Gamma_2^\star/\sqrt{\ln 2}}\right)^2\right] \\ &\cdot\ \mathrm{erfc}\left(-i\frac{(\omega - \omega_s) + i\gamma}{2\Gamma_2^\star/\sqrt{\ln 2}}\right). \end{aligned} \tag{3.27}$$

The emergence of a mode splitting is a good indication of strong coupling. However, the distance of the peaks and their shape can be quite different depending on the frequency distribution of the spins. Figure 3.4 compares the power spectrum $|S_{21}|^2$ of a CPW resonator coupled to a spin ensemble with a Lorentzian and a Gaussian line shape. The collective coupling strength ($v_N/2\pi = 12\,\mathrm{MHz}$) is kept fixed as well as the cavity decay rate ($\kappa/2\pi = 5\,\mathrm{MHz}$) and the total inhomogeneous linewidth ($\Gamma_2^\star/2\pi = 4\,\mathrm{MHz}$). From left to right, the fraction of the linewidth of spins following a Gaussian distribution is increased. This is achieved by using a Voigt profile, which is a convolution of Gaussian and Lorentzian distributions. The $W(\omega)$ function for this case turns out to be given by $W_{\mathrm{Gauss}}(\omega)$ and replacing $i\gamma \rightarrow i\gamma + i\Gamma_{\mathrm{Lorentzian}}$.

As one can see, the shape of the inhomogeneous spin distribution clearly influences the spectrum. Although the coupling and absolute linewidth do not change, the peaks of the mode splitting appear further apart and narrower in the case of a Gaussian inhomogeneous distribution. Specific distributions give rise to new effects such as *cavity protection* [Din+11], where the cavity suppresses the decay of excitations to the dark modes. This is the case for inhomogeneous spin distributions $\rho(\omega)$ decaying faster than ω^{-2}. The interaction of inhomogeneously broadened spin ensembles with a microwave resonator is an active field of research and Refs. [Put+14; Kri+14] are recommended for further reading. Most of the time, however, we will assume a Lorentzian distribution for the sake of simplicity.

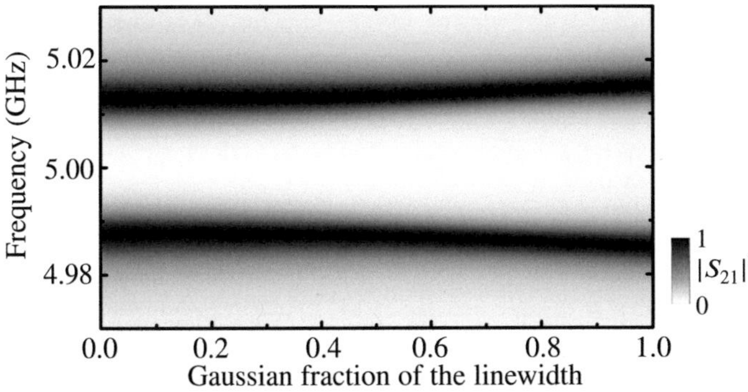

Figure 3.4: Power spectrum as a function of probe frequency for different inhomogeneous spin line shapes from purely Lorentzian (left) to purely Gaussian (right).

Another convenient type of probing a whole array of resonators is the notch type configuration (Sec. 3.1.3). The derivation of the S_{21} transmission is analogous. Neglecting the homogeneous broadening and assuming a Lorentzian inhomogeneous broadening, one obtains [Sch+10]

$$S_{21} = 1 - \frac{\kappa_c}{i(\omega - \omega_c) - (\kappa_c + \kappa_i) + \frac{\upsilon_n^2}{i(\omega - \omega_s) - \Gamma_2^\star}} \,. \tag{3.28}$$

Here, $2\kappa_c$ and $2\kappa_i$ denote the coupling and internal decay rates of the resonator.

3.2.5 Temperature dependence of the collective coupling

The thermal population of the energy levels of a paramagnetic spin 1/2 system is given by the Boltzmann distribution

$$P(|x\rangle) = \frac{\exp\left(-\frac{E_x}{k_B T}\right)}{\exp\left(-\frac{E_\uparrow}{k_B T}\right) + \exp\left(-\frac{E_\downarrow}{k_B T}\right)} \,, x \in \{\uparrow, \downarrow\} \,, \tag{3.29}$$

with the energy of ground and excited state being $E_\downarrow = -\frac{1}{2} g \mu_B B$ and $E_\uparrow = \frac{1}{2} g \mu_B B$, respectively. The coupling strength υ_N is proportional to the square root of the net magnetic moment

M_z, which depends on the difference in the state population, $\upsilon_N = \upsilon_0\sqrt{N\left[P(|\downarrow\rangle - P(|\uparrow\rangle)\right]}$. After inserting Eq. (3.29), the expression is simplified using $\tanh(x) = (e^x - e^{-x})/(e^x + e^{-x})$

$$\upsilon_N = \upsilon_0\sqrt{\tanh\left(\frac{g\mu_0 B}{2k_B T}\right)}\,. \tag{3.30}$$

This expression only holds true for a perfect paramagnet, where the spin-spin interaction is negligible.

3.3 Measurement of the spin lattice relaxation

Direct microwave absorption

A convenient way to study the spin lattice relaxation (SLR) is by measuring the direct microwave absorption signal of the spin ensemble coupled to a microwave transmission line [Pro+13; Cla+13]. When the frequency of the incident microwave signal is in resonance with the spins, it gets attenuated because part of the microwave power is absorbed by the spins. In a typical setup, the $S_{21}(f)$ transmission is measured such that the coupling to the spins appears as a dip in the power spectrum $|S_{21}(f)|^2$. For very low microwave powers, i.e. no saturation of the ensemble, the magnitude of the absorption dip is proportional to the population difference of spins in the ground and in the excited state $\Delta N = N_1 - N_2$, which is proportional to the magnetization of the ensemble M_z. In order to measure the spin relaxation time, a strong and resonant microwave pulse is applied which saturates the spin system $\Delta N = 0$. Then, the $|S_{21}|^2$ transmission is measured at the same frequency but at low microwave power and recorded over time [SE01]. After the pulse, the ensemble relaxes back to its equilibrium level, which is $N_2/N_1 = \exp\left(-\hbar\omega_s/k_B T\right) \sim 10^{-5}$ at 20 mK temperature. The spin lattice relaxation signal is given by the $|S_{21}|^2(t)$ signal at time t compared with the transmission magnitude right after the saturation pulse $|S_{21}|^2(0)$

$$|S_{21}|^2(0) - |S_{21}|^2(t) \propto N_1 - N_2 \propto M_z\,. \tag{3.31}$$

At low temperatures $\hbar\omega \gg k_B T$, the SLR of paramagnetic ions in crystals is dominated by a direct process and the SLR time T_1 is given by

$$\frac{1}{T_1} = R_d\,(\hbar\omega)^5 \coth\left(\frac{\hbar\omega}{2k_B T}\right), \tag{3.32}$$

where the parameter R_d describes the induced emission and absorption of photons, see Sec. 2.3.5.

Spin ensemble coupled to a resonator

In hybrid circuit QED experiments, it is not always possible to detect a strong direct microwave absorption signal due to low spin concentrations. Nonetheless, in such a situation, it is still possible to detect the spin ensemble by employing a resonator due to the $\sqrt{N}$ enhancement of the coupling strength (see Sec. 3.2.2). The amplitude transmission spectrum $|S_{21}|$ of a resonator coupled to a transmission line shows a dip at the resonance frequency. If a spin ensemble is in resonance with the resonator, the dip gets shallower because a part of the power is absorbed by the spin ensemble. The amplitude of the dip in the $|S_{21}|$ spectrum is proportional to

$$|S_{21}| \propto 1 - \frac{\kappa_c}{\kappa} = 1 - \frac{\kappa_c}{\kappa_l} \frac{1}{1 + v_N^2 / \Gamma_2^\star \kappa_l}, \tag{3.33}$$

where κ_l and κ_c describe the loaded and coupling dissipation rates of the resonator. The collective coupling strength is given by v_N and the inhomogeneous spin linewidth is denoted by $\Gamma_2^\star$. The quantity $v_N^2 / \Gamma_2^\star \kappa_0$ in the denominator is called cooperativity C. In the case of weak coupling, C is smaller than 1 whereas in the case of strong coupling, it is larger than 1. For very weakly coupled spin ensembles, i.e. $C < 0.1$, Eq. (3.33) can be approximated by performing a Taylor expansion with respect to C such that $|S_{21}| \propto 1 - \kappa_c / \kappa_l + \kappa_c v_N^2 / \Gamma_2^\star \kappa_l^2$. The second part of the equation contains the collective coupling squared, which is directly proportional to the number of spins, i.e. the population difference $N_1 - N_2$.

The protocol for determining the SLR time is similar to the case of direct microwave absorption. First, a strong pulse saturates the ensemble. According to Eq. (3.33), the dip of the resonator gets deeper because the collective coupling strength is zero right after the pulse. After the pulse, the resonator is monitored at very low power to avoid further saturation of the spin ensemble. While the ensemble relaxes back to its equilibrium state, the dip of the resonator gets shallower again. In the case of very weak coupling $C < 0.1$, it is possible to fit the time dependence of the amplitude of the resonator dip directly with an exponential decay. Note that in contrast to direct absorption one analyzes the amplitude instead of the power. It turns out that the fit yields also reasonable values for $C > 0.1$, because the T_1 time of the exponential decay is mostly influenced at the beginning, where C is still very small. For arbitrary C values, one has to fit the full model

$$|S_{21}| = A \left[1 - B \frac{1}{1 + C \left[1 - \exp\left(-\frac{t - t_0}{T_1} \right) \right]} \right]. \tag{3.34}$$

The amplitude A corresponds to the baseline of the transmission spectrum far away from the resonance, B is given by κ_c / κ_l, C is the cooperativity in equilibrium, t_0 denotes the time directly after the saturation pulse and T_1 is the SLR time. The derivation for a transmission line resonator goes analog and Eq. (3.34) transforms to

$$|S_{21}| = B \frac{1}{1 + C \left[1 - \exp\left(-\frac{t - t_0}{T_1} \right) \right]}. \tag{3.35}$$

3.4 Microwave-optical conversion

Erbium doped crystals are well suited for implementing a reversible single photon microwave to telecom C-band converter. The biggest challenge for implementing an efficient scheme is imposed by the inhomogeneous broadening of both the optical and microwave transitions. In the following, a conversion protocol with a theoretical efficiency of 90% is presented [OBr+14]. It employs spectral hole burning and controlled reversible inhomogeneous broadening (CRIB) [Hét+08; Spa+13] in conjunction with π-pulses to achieve a large bandwidth and a refocusing of the dephasing induced by the inhomogeneous broadening.

The system consists of four levels and is shown in Fig. 3.5(a). The lowest two states form the spin ensemble ($|\downarrow\rangle$, $|\uparrow\rangle$) and the ground and the topmost state form the optical ensemble ($|\downarrow\rangle$, $|e\rangle$). The fourth state is an auxiliary level ($|x\rangle$). First, the system starts in the ground state and by optical pumping all the population is transferred to the auxiliary state. Then, a narrow band laser selectively transfers population back to the ground state such that a very narrow frequency ensemble is formed. The system preparation finalizes by applying a gradient magnetic field, which broadens the ensemble in frequency space.

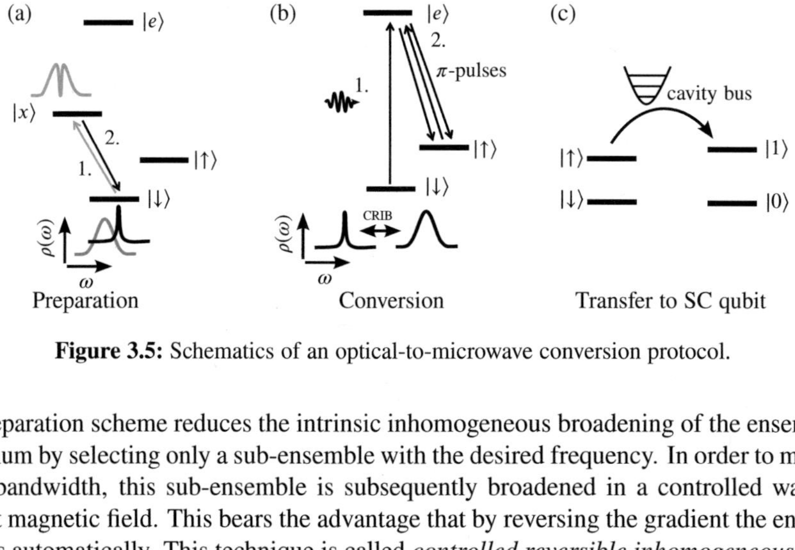

Figure 3.5: Schematics of an optical-to-microwave conversion protocol.

This preparation scheme reduces the intrinsic inhomogeneous broadening of the ensemble to a minimum by selecting only a sub-ensemble with the desired frequency. In order to maintain a high bandwidth, this sub-ensemble is subsequently broadened in a controlled way by a gradient magnetic field. This bears the advantage that by reversing the gradient the ensemble rephases automatically. This technique is called *controlled reversible inhomogeneous broadening* (CRIB).

In Fig. 3.5(b), an incident optical photon (flying qubit) is absorbed and its coherence is stored in the coefficients of the $|\downarrow\rangle$ and $|e\rangle$ levels. The aim of the following procedure is to transform this coherence to a SC qubit (stationary qubit). After a certain time τ_1, the coherence is transfered from the $|e\rangle$ to the $|\uparrow\rangle$ state by an optical π-pulse. Then, one waits for another time τ_2 after which the population is transfered back to the $|e\rangle$ state and the gradient magnetic field is reversed (CRIB technique). The reversal of the magnetic field effectively inverts the time evolution such that at time $2\tau_1$, a photon echo should be emitted. In this case, however, no photon echo will occur due to the dephasing of the spin state. At this moment, another optical π-pulse transforms the coherence back to the spin state $|\uparrow\rangle$, where the rephasing continues. In Fig. 3.5(c), the system is again in phase and a tunable microwave cavity similar to the one employed in Ref. [Kub+11] is brought into resonance for the time of a Rabi swap. The excitation is now in the cavity which is subsequently tuned into resonance with a qubit where the excitation gets transferred to. Note that the procedure is reversible.

An alternative to CRIB would be to employ an atomic frequency comb (AFC) [Afz+09], which rephases itself after a time determined by the inverse of the comb spacing.

4 Experimental setup

This chapter describes the general experimental setup, which is used throughout the theses. It is capable of performing both microwave (in the 3-12 GHz range) and optical ($\lambda \sim 1.54\,\mu m$) spectroscopy at milli-Kelvin temperatures.

4.1 Cryostat

All investigations were performed inside a cryogen free Bluefors BF-LD-250 dilution refrigerator with a base temperature of 20 mK and anti-reflection coated ($\lambda \approx 1.54\,\mu\text{m}$) windows at two opposite sides (Sapphire and Spectrosil 2000, UQG Optics Ltd) allowing for optical spectroscopy at milli-Kelvin temperatures. The refrigerator is cooled down to approximately 3.6 K by a cryomech pulse tube cooler before the dilution unit cools down further to base temperature. The ^{3}He / ^{4}He mixture gets liquefied by a Joule-Thomson valve. The cryostat is separated into several temperature stages. The 60 K stage and the 4 K stage are cooled by the pulse tube, the still stage and the mixing chamber (MC) are cooled by the dilution circuit.

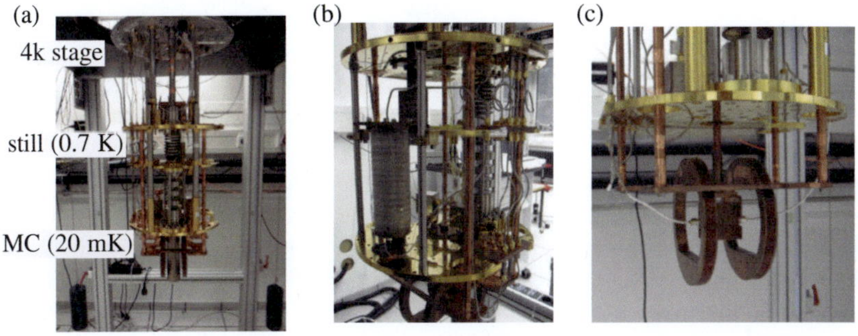

Figure 4.1: (a) Lower levels of the cryostat, the top silvery plate is the 4 K stage with typical temperatures between 3.5 and 3.8 K. All HEMT amplifiers are mounted there. The golden plate in the middle is the still stage and the bottom plate is the mixing chamber stage with a base temperature of 20 mK. (b) Solenoid coil, mounted at the still level. The sample is anchored at the MC stage and enters the solenoid coil from the bottom via a metallic finger. (c) The Helmholtz coil pair is supported by long copper tubes, which are fixed at the still stage. The sample hangs in the center, thermally anchored to the MC plate. In the picture a 3D cavity is installed.

Figure 4.1(a) shows a photograph of the lower levels of the cryostat. All cryogenic HEMT[1] amplifiers are installed at the 4 K stage and all samples are thermally anchored to the MC stage. At the still level, a self-built superconducting solenoid coil and a Helmholtz coil pair are mounted, see Figs. 4.1(b) and 4.1(c), respectively. The solenoid coil has a maximum field of 370 mT and is used for on-chip electron spin resonance (ESR) experiments in order to investigate the spin ensembles. The Helmholtz coil pair has a maximum field of 275 mT and the setup is used for both ESR and optical studies. The symmetry axis of the Helmholtz coil pair is parallel to the optical axis defined by the windows of the cryostat. The sample space can be used with standard on-chip ESR sample holders as well as with 3D resonators and shielded qubit sample holders. In addition, it is possible to control the two coils of the Helmholtz coil pair individually in order to apply magnetic field gradients at the sample. Figure 4.2(a) shows the circuit diagram for powering the solenoid and the Helmholtz coils and Fig. 4.2(b) displays the wiring inside the cryostat at the 4 K stage. The currents for the coils are supplied either by two Tektronix PWS4205 power supplies connected in parallel, or a Keithley SMU2602A SourceMeter and a Yokogawa GS820 source measure unit connected in parallel, too. From the 60 K stage to the 4 K stage, four high-T_c superconductor cables are used. From the 4 K stage, superconducting NbTi wires[2] are employed. In the following sections, the general setup for microwave and optical spectroscopy is described.

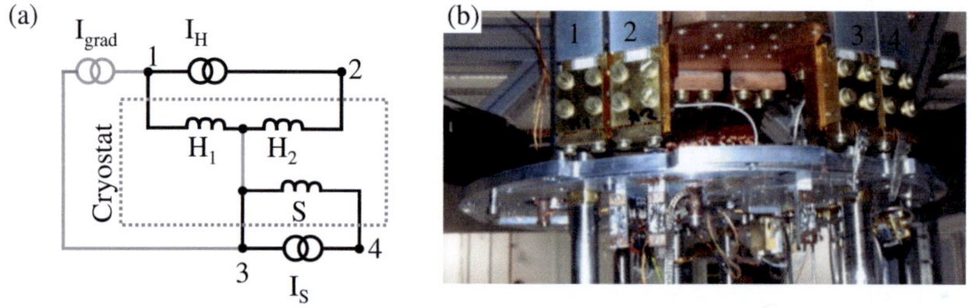

Figure 4.2: (a) Circuit layout for powering the Helmholtz (H_1, H_2) and the solenoid (S) coils. The part of the circuit, which is inside the cryostat, is marked by the dotted rectangle and the connections into the cryostat are labeled by numbers 1-4. By adding the grey part to the circuit including an additional current supply, gradients can be applied in the Helmholtz coil setup in order to artificially enlarge the inhomogeneous broadening of the spin ensemble. (b) Picture of the 4 K stage with the wiring for the coils biases. Four high-T_c cables conduct the current from the 60 K to the 4 K stage where wiring is continued using superconducting NbTi wires. The connections shown in (a) are done at the three posts anchored below the 4 K plate and superconducting tin-solder provides good electrical contact.

[1] A *high-electron-mobility transistor* (HEMT) is a heterostructure field-effect transistor, which is able to operate at higher frequencies than ordinary transistors.

[2] Single filament NbTi superconductor with copper cladding, diameter ≈ 0.152 mm plus 0.178 mm insulation, (type T48B-M) from Superconductor Inc.

4.2 Microwave spectroscopy

The cryostat is equipped with seven microwave lines, of which four are used for input and three for output. All cables from the 60 K stage down to the MC are stainless steel coaxial cables in order to minimize the thermal conductivity between the stages while maintaining a good electrical conductivity. The cables for the part from the 60 K stage to room temperature are made from copper for the three output lines and stainless steel for the four input lines. Most of the on-chip ESR investigations are performed in the single microwave photon regime. In order to maintain a good signal to noise ratio, the incident microwave probe signal is attenuated at several temperature stages by Inmet-Aeroflex AH9026-XX attenuators, see Fig. 4.3. After the signal has passed the sample, it is guided through a shielded cryogenic microwave circulator (Quinstar CTH0408KCS) where the third port has been terminated by a 50 Ω resistor, which shields the sample from the backaction of the cryogenic HEMT amplifier.

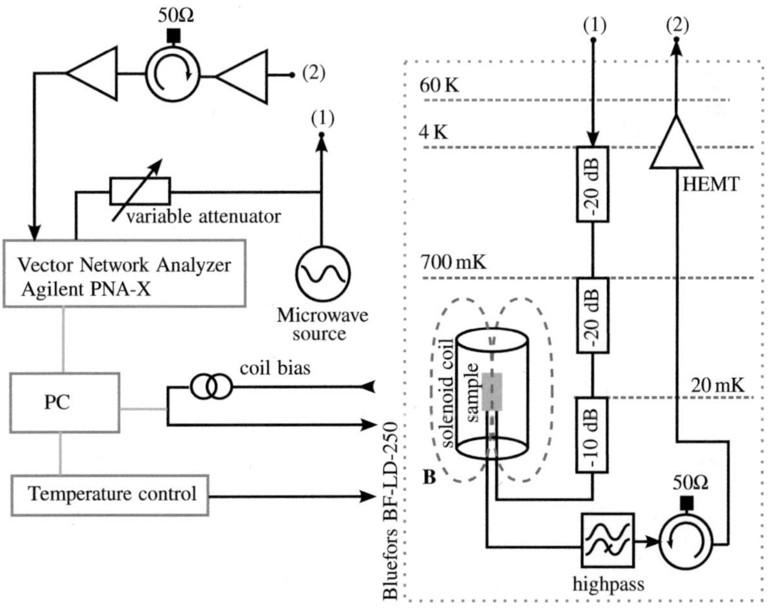

Figure 4.3: Setup for continuous wave microwave on-chip spectroscopy. The sample is placed inside a solenoid coil and cooled down to 20 mK. Typical probing powers at the sample range from −120 dBm to −140 dBm. Cold attenuators, a cryogenic circulator and HEMT ensure a good signal-to-noise ratio down to the single photon regime.

The refrigerator contains three HEMT amplifiers, which are all anchored to the 4 K stage. Two of them are used in this thesis: LNF-LNC4_8A featuring a bandwidth of 4-8 Ghz with 40 dB gain and 2.6 K noise temperature, and LNF-LNC4_8A having a 6-20 GHz bandwidth with 31 dB gain and a noise temperature of 9.5 K. At room temperature, the signal is further amplified by two Minicircuits ZVA-183-S+ amplifiers. These amplifiers are separated by a circulator (Pasternack PE8402), where again the third port has been terminated by a 50 Ω resistor. Depending on the scope of the setup, additional filters are inserted. Typically, the on-chip ESR experiments presented in this theses use a 3-7 GHz bandpass filter between the sample and the circulator. The continuous wave microwave spectroscopy is carried out by measuring the $S_{21}(f)$ transmission with an Agilent Technologies N5241A PNA-X Vector Network Analyzer and sweeping the magnetic field of either the solenoid or the Helmholtz coil.

The superconducting Nb resonators used in the on-chip ESR spectroscopy were fabricated a the IMS[3] by Dr. Stefan Wünsch. Figure 4.4(a) shows a lumped element (LE) resonator of dimensions 0.6×0.6 mm^2 fabricated at the IMS [Wue+11]. These LE resonators were arranged in arrays of three to nine resonators per chip connected by a common 50 Ω transmission line. The resonators in an array have different resonance frequencies covering the band from 4 to 5 Ghz to allow multiplexed readout [Jer+12; Che+12]. They possess loaded and internal single photon quality factors of $\sim 10^3$ and $\sim 10^4$, respectively.

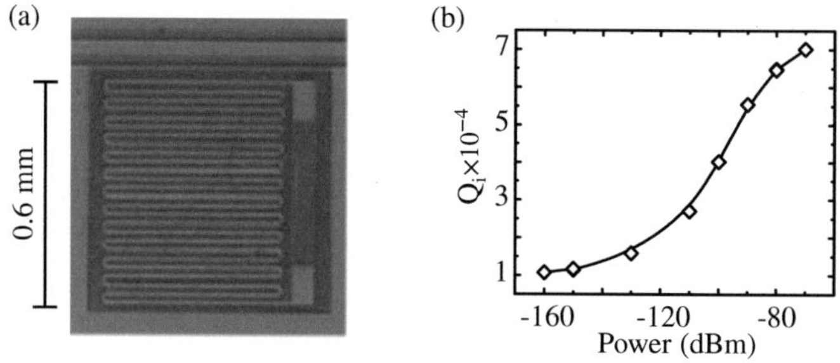

Figure 4.4: (a) Optical micrograph of a LE resonator. (b) Power dependence of the internal quality factor. The solid line is a guide to the eye.

Figure 4.4(b) presents the internal quality factor of one LE resonator as a function of power. At high power, the internal quality factor Q_i is seven times larger than at low microwave power. This dependency results from the coupling to parasitic two level fluctuators caused by impurities in the superconductor or substrate [Gao+07; Gao+08b; Gao+08a; Wan+09]. The saturation of those defect states at high power effectively reduces the overall photon loss. The quality factors were determined using the algorithm presented in Ref. [Pro+14b]. With regard to the coupling strength between the resonators and the spin ensembles, LE resonators are comparable to conventional coplanar waveguide (CPW) resonators. However, LE resonators are more compact and can easily be arranged in multiplexed arrays to study several samples at the same time. Furthermore, they outperform CPW resonators in large magnetic fields, which are needed for the investigations of Er:YSO at small DC g-factors. We found that our LE resonators can withstand an in-plane magnetic field up to 340 mT without degrading significantly. In a few experiments, other types of microwave resonators were used; a LE

[3] Institut für Mikro- und Nanoelektronische Systeme, Karlsruhe Institute of Technology, D-76189 Karlsruhe, Germany

resonator fabricated on a Nd:YAP crystal (Sec. 5.2), a copper CPW resonator (Sec. 6) and a 3D cooper cavity (Sec. 5.4). They will be discussed in the respective experimental sections.

4.3 Optical spectroscopy

For the optical spectroscopy at 1.54 µm wavelength, a Toptica DL 100 grating stabilized single mode diode laser is used. Its wavelength is adjusted manually by a screw and a fine adjustment including periodic and triggered frequency sweeps are possible via a piezo control. An optical isolator (Thorlabs IO-5-1550-HP) at the output of the laser prevents damage by backwards propagating beams. The wavelength of the laser is measured by an Ångstrom high-Finesse WS6 wavemeter. A Brimrose acousto-optical modulator allows us to modulate the intensity of the beam on short timescales ($\sim 50\,\text{ns}$) with a modulation signal provided by a Tektronix AWG520 arbitrary waveform generator.

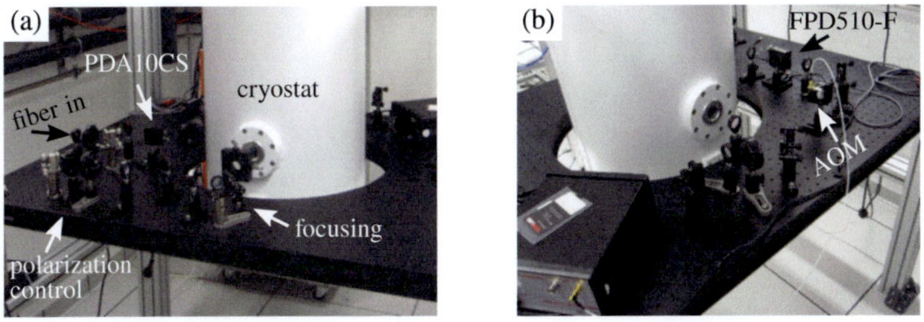

Figure 4.5: Second small optical table at the cryostat. (a) The laser beam is coupled out of the fiber, linear polarized, widened to approximately 1 cm diameter and finally focused into the fridge. In the background, the Thorlabs PDA10CS photodetector is seen. (b) Optical output of the cryostat. The beam is focused on the AOM switching it between two photo detector.

The optical probe signal is prepared on a standard optical table. Then, it is coupled into a polarization maintaining fiber and guided to a smaller optical table, which surrounds the Bluefors BF-LD-250 dillution fridge, see Fig. 4.5. On this small table, the light is coupled out of the fiber and sent through a polarization control stage before it is focused onto the sample inside the cryostat. The focusing is realized with three lenses: First, a concave ($f = -7.5\,\text{cm}$) and a convex lens ($f = 25\,\text{cm}$) widen the parallel beam to approximately 1 cm diameter. Second, a convex lens ($f = 30\,\text{cm}$) focuses the beam onto the sample. Using Gaussian beam propagation, the beam waist at the sample is calculated to be approximately 60 µm. For a 1 mW laser beam, this corresponds to an intensity of about $350\,\text{kW/m}^2$ at the sample. The corresponding average electric and magnetic fields are 510 V/m and 1.7 µT, respectively.

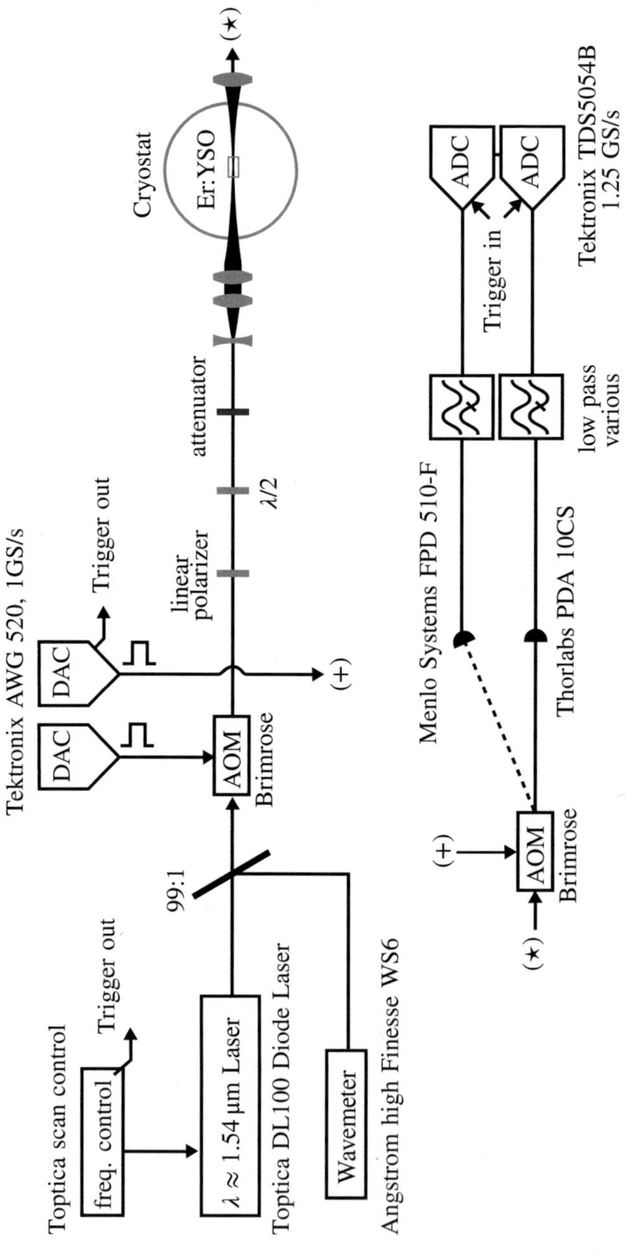

Figure 4.6: Setup for optical spectroscopy.

The transmitted light exits the cryostat at the opposite side and is guided through an additional acousto-optical modulator. It allows us to switch quickly between two photodetectors, a slower Thorlabs PDA10CS switchable gain detector and a fast Menlo Systems FPD510-F photodetector with 250 MHz bandwidth. The voltage signals from the detectors are recorded by a Tektronix 5054B oscilloscope.

4.4 Rare earth doped crystals

Most of the rare earth (RE) doped crystals were grown by Scientific materials Inc.[4] using the Czochralski method. The company provides the dopant concentration c_d in relative atomic percent, that is, $c_d = N_d/N_{\mathrm{sub}}$. Here, N_d denotes the number of RE dopant atoms and N_{sub} denotes the number of atoms in the bulk, which can be replaced by the dopant (Y in our case). Typical concentrations range from 10 to 1000 ppm. In this thesis, the host materials Y_2SiO_5 and $YAlO_3$ doped with the rare earth elements Er and Nd are investigated. The Nd doped $YAlO_3$ sample has been provided by the MTI company[5].

In addition to bulk doped crystals, our collaborators from RUB Bochum performed focused ion beam irradiation with Er^{3+} ions onto undoped Y_2SiO_5 crystals. The ESR properties of those surface doped samples are investigated in Sec. 5.3.

[4] `http://www.scientificmaterials.com`

[5] `http://www.mtixtl.com/yalo3nd-doped10010x10x05mm1sidepolished.aspx`

5 Anisotropic coupling of microwave resonators to rare earth spin ensembles

This chapter presents the results on interfacing microwave resonators to rare earth spin ensembles (Er, Nd) in the few photon limit. The collective coupling strength and the inhomogeneous spin linewidth is analyzed with respect to temperature and magnetic anisotropy.

5.1 Anisotropic coupling to rare earth spin ensembles doped into Y_2SiO_5

In the following investigations, 200 ppm erbium doped Y_2SiO_5 (Er:YSO) crystals are mounted onto a superconducting (SC) niobium chip containing 9 lumped element (LE) resonators occupying the band between 4.5 and 5.2 GHz. The LE resonators are connected to a 50 Ω transmission line and possess loaded and intrinsic quality factors of $Q_l \simeq 10^3$ and $Q_i \simeq 10^4$, respectively. For tuning the spins, the static magnetic field is applied in parallel to the chip's surface to avoid degradation of the SC resonators by induced screening currents. For our setup, we find that fields up to 340 mT have no significant influence on the LE resonator properties.

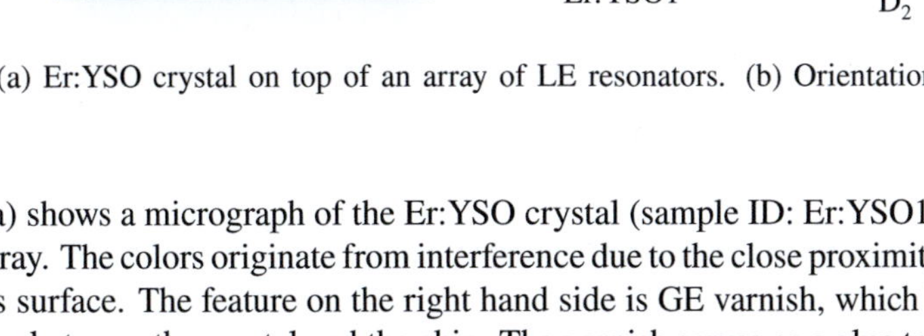

Figure 5.1: (a) Er:YSO crystal on top of an array of LE resonators. (b) Orientation of the axes of the crystal.

Figure 5.1(a) shows a micrograph of the Er:YSO crystal (sample ID: Er:YSO1) on top of the resonator array. The colors originate from interference due to the close proximity of the crystal to the chip's surface. The feature on the right hand side is GE varnish, which has penetrated the small gap between the crystal and the chip. The varnish serves as a glue to fix the crystal on the SC chip. The LE resonators covered by the GE glue possess much lower quality factors, presumably originating from dielectric losses and parasitic two level systems in the glue. It is important that the glue is applied such that the crystal experiences no mechanical stress when it is cooled down because this would increase the inhomogeneous spin linewidth. Figure 5.1(b) illustrates the cut of the crystal with respect to the crystal axes. By varying the direction of the magnetic DC field with respect to the crystal axes, the magnetic anisotropy of the Er^{3+} ions is studied and the collective coupling strength to the LE resonator as well as the inhomogeneous spin linewidth are extracted for each direction.

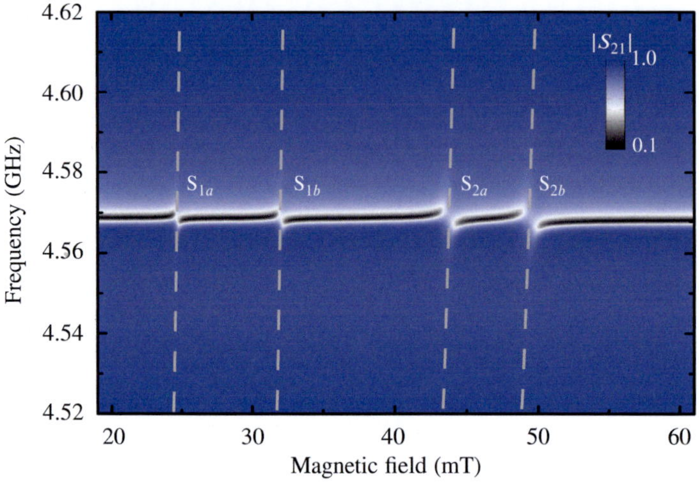

Figure 5.2: On-chip ESR spectrum of sample Er:YSO1 coupled to a LE resonator. The orientation of the crystal's b-axis with respect to the DC magnetic field is $\theta = 81.3°$.

Figure 5.2 shows a typical on-chip ESR spectrum of the Er:YSO1 crystal. The transmitted microwave amplitude $|S_{21}|$ is plotted in color vs. frequency and magnetic field. The spectrum was recorded in the few photon regime at a temperature of 22 mK. The black line, which corresponds to the resonator's absorption dip, experiences four dispersive frequency shifts. There, the resonator couples to the spin system and the corresponding energy loss manifests itself by a reduced quality factor. Theses are caused by the coupling of the resonator to the electronic spin transitions of the four sub-ensembles of the Er:YSO1 spin ensemble.

The spectrum is analyzed in the following way: First, the resonance frequency of the LE resonator is extracted as a function of the magnetic field by fitting the spectrum using the circle fit method [Pro+14b]. Second, the dispersive shifts of the resonance frequency are fitted to the weak coupling theory presented in Sec. 3.2.3 in order to extract the collective coupling strength υ_N, the inhomogeneous spin linewidth $\Gamma_2^\star$ and the DC g-factor g_{DC}. The results are presented in the following table.

Crystal	Er:YSO1 (200 ppm, $\theta = 81.3°$)			
transition	B_{DC} (mT)	g-factor	$\upsilon/2\pi$ (MHz)	$\Gamma_2^\star/2\pi$ (MHz)
S_{1a}	24.1	13.6	6.7	25.5
S_{1b}	30.5	10.7	7.0	25.5
S_{2a}	40.9	8.0	10.2	28.8
S_{2b}	46.0	7.1	11.0	30.4

Here, the subscripts of the labels $S_{...}$ indicate the electronic transitions of the erbium ions at site 1 or 2 and magnetic subclass *a* or *b*. For all four transitions, we observe a collective coupling strength that is smaller than the inhomogeneous linewidth. Transitions with smaller DC g-factors possess a larger collective coupling strength but also a larger inhomogeneous spin linewidth. The increase of the linewidth can be explained by inhomogeneities in the magnetic DC field. The erbium doped crystal is glued onto a superconductor which is a perfect diamagnet. The homogeneous DC field, which is responsible for the Zeeman splitting of the spin transitions, should be aligned parallel to the surface of the superconducting LE resonator chip. However, the orientation of the magnetic field is not perfect and the residual perpendicular component gives rise to counteracting fields created by the superconductor. This inhomogeneity broadens the distributions of spin transition frequencies. By employing a normal metal resonator, one can avoid this effect, but normal metal resonators have low quality factors due to their the finite electrical resistivity, see Sec. 6.1. By linear extrapolation of the linewidth to zero magnetic field, we obtain an intrinsic inhomogeneous spin linewidth of $\Gamma_2^\star/2\pi \sim 19 \pm 1.4$ MHz. Note, this value sets a lower limit for the collective coupling strength for obtaining strong coherent coupling, i.e. $\upsilon_N/2\pi \gg 19$ MHz. The increase of the collective coupling strength has a different cause. The magnetic anisotropy of the erbium ions in YSO is described by a g-tensor **g**. Therefore, the effective magnetic moment of the erbium spin depends on the direction. The full Hamiltonian is given by

$$H = \mu_B \mathbf{B}_{DC} \cdot \mathbf{g} \cdot \mathbf{S} + \mu_B \mathbf{B}_{AC} \cos \omega t \cdot \mathbf{g} \cdot \mathbf{S} \, . \tag{5.1}$$

The first term describes the anisotropic Zeeman interaction of the spin with the static magnetic field and the second term specifies the anisotropic coupling to the oscillating magnetic field. In a typical experiment, $\mathbf{B}_{DC}$ and $\mathbf{B}_{AC}$ are perpendicular to each other so that they experience different effective g-factors. For materials with axial symmetry, small DC g-factors are accompanied by large AC g-factors, see Sec. 2.3.3, Fig. 2.2. Thus, it is necessary to explore the anisotropic coupling in order to enter the strong coupling regime, where $\upsilon_N \gg \Gamma_2^\star, \kappa_l$.

Furthermore, it is crucial to investigate the temperature dependence of the collective coupling strength and the inhomogeneous spin linewidth. The temperature dependence can help to identify the given spin ensemble as a paramagnet, i.e. an ensemble of uncoupled spins (on the timescales of interest). In order to do this, spectra similar to Fig. 5.2 were recorded at different temperatures and the extracted couplings and linewidths for all four transitions are presented in Fig. 5.3.

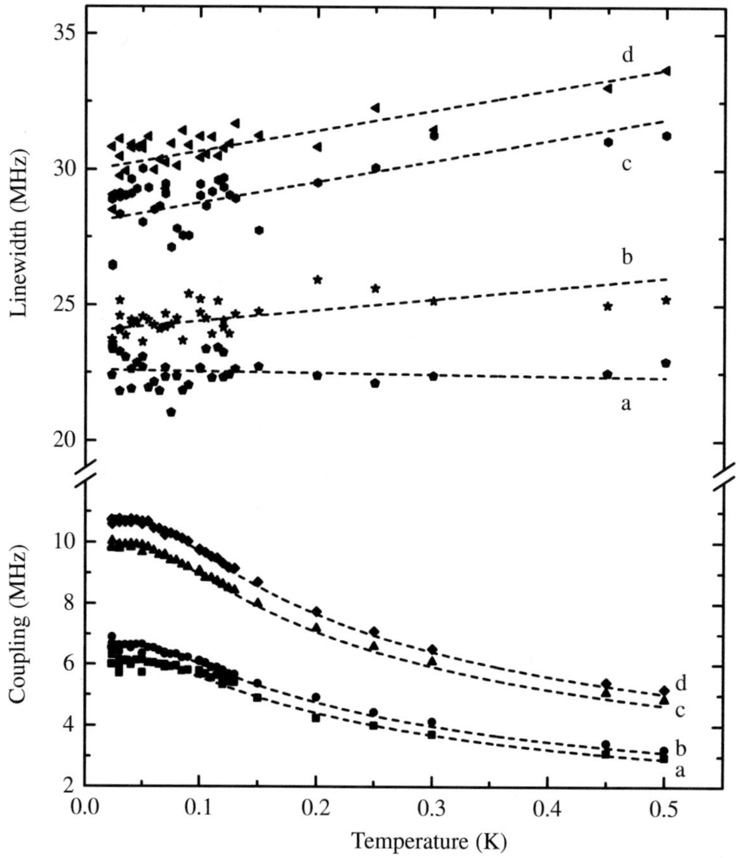

Figure 5.3: Temperature dependence of the collective coupling strength and the inhomogeneous spin linewidth. The error bars are omitted for better readability (linewidth: $< \pm 1$ MHz, coupling: $< \pm 0.06$ MHz). The coupling strength fits well to the theory of paramagnetic ions in solids (lower dashed line). The larger error in determining the linewidth only allows for a qualitative analysis. The upper dashed lines show the general trend that the linewidth increases with increasing temperature. The labels a-d numerate the transitions in the spectrum starting with the lowest field transition.

The temperature dependence of the collective coupling strength was fitted to the standard theory of paramagnetic ions in crystals, see Sec. 3.2.5,

$$\upsilon_N = \upsilon_0 \sqrt{\tanh\left(\frac{\hbar\omega}{2k_B T}\right)}, \tag{5.2}$$

with $\omega = 2\pi \times 4567.5$ MHz. Figure 5.3 shows a nice agreement between theory and experiment for all four sub-ensembles, and confirms the paramagnetic properties of the 200 ppm Er^{3+} doped Er:YSO1 sample. The determination of the inhomogeneous spin linewidth from the fit to the spectrum is subject to a larger error and the data allow only for a qualitative analysis.

The general trend shows that the linewidth increases with increasing temperature, except for the lowest spin transition (see Fig. 5.3a). However, the large spread of values (±0.5 to ±1 MHz), does not allow for further conclusions at this point.

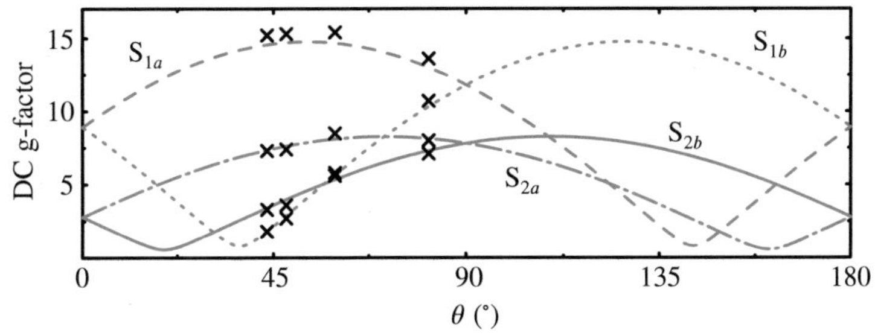

Figure 5.4: Simulation of the DC g-factors of sample Er:YSO1 as a function of θ for fixed $\phi = 111°$. The black crosses indicate the measured DC g-factors.

Sample Er:YSO1 was studied in four different orientations. For each orientation and for all four sub-ensembles, collective coupling strength υ_N, inhomogeneous spin linewidth $\Gamma_2^\star$ and DC g-factor were extracted. In order to understand anisotropic behavior, the data were compared to a simulation of the system Hamiltonian (see Sec. 2.2) using the Matlab package EASYSPIN [SS06] and the g-tensors provided in Ref. [Sun+08]. Figure 5.4 presents the measured data on top of a simulation of the DC g-factors. The simulation yields the best agreement with the data for $\phi = 111°$, which is within reasonable agreement with the crystal cut provided by our supplier (Fig. 5.1(b)) $60° \sim 69° = 180° - \phi$.

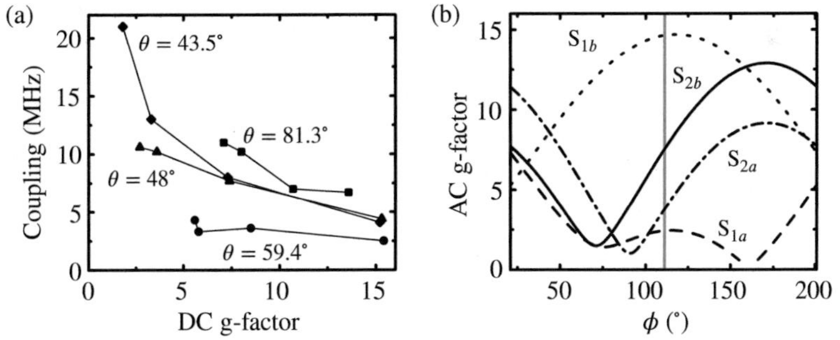

Figure 5.5: Anisotropic coupling. (a) Coupling vs. DC g-factor. For almost all samples, the coupling scales inversely to the DC g-factor. (b) Simulated angular dependence of the AC g-factor for $\theta = 44°$. The grey line denotes the mean orientation. The coupling measured with on-chip ESR represents an average value with respect to the AC g-factor anisotropy due to the inhomogeneous AC field of the LE resonator.

All samples show the general trend that transitions with smaller DC g-factors possess larger coupling strengths, see Fig. 5.5(a). The reason for the change in coupling strength is also attributed to the magnetic anisotropy. The AC magnetic field of the LE resonator is oriented in the plane perpendicular to the static magnetic field. However, it is still very inhomogeneous in this plane instead of pointing in only one direction. The coupling strength of an individual spin to the LE resonator depends on its effective magnetic moment $\mu_{\text{eff}} = \mu_B g_{\text{AC}}$. Figure 5.5(b) shows the angular dependence of the so-called AC g-factor for an orientation of $\theta = 44°$ of the DC field. Thus, in order to calculate the collective coupling strength, one has to take into account the angular variations of the vacuum magnetic field $\mathbf{B}_{\text{vac}}(\mathbf{r})$ and the g-factor

$$\upsilon_N^2 = \int \rho(\mathbf{r})\,|\upsilon(\mathbf{r})|^2\,\mathrm{d}\mathbf{r}\,, \tag{5.3}$$

where $\upsilon(\mathbf{r}) \propto \mathbf{B}_{AC}\mathbf{g}\,\mathbf{S}$ and $\rho(\mathbf{r})$ denotes the density of spins, which is typically assumed to be homogeneous.

Although the collective coupling υ_N increases for smaller DC g-factors, coherent strong coupling is not possible because the inhomogeneous spin linewidth is always larger than υ_N. Therefore, a different orientation was chosen. For this, a new crystal (sample ID: Er:YSO2) was employed with the same dopant concentration as Er:YSO1 and a 30 degrees rotated D_1 axis such that the axis coincides with one of the crystal edges, see Fig. 5.6(ii). In this experiment both Er:YSO1 and Er:YSO2 were measured simultaneously, each crystal covering different LE resonators of the array. The crystals were aligned such that the DC magnetic field forms angles of approximately 43.5° (Er:YSO1) and 48° (Er:YSO2) with the crystallographic b axis.

Figure 5.6 presents the on-chip ESR spectrum. The resonators covered by erbium doped crystals show again a number of dispersive shifts. The dotted and dashed lines indicate the 4 sub-ensembles of Er:YSO1 and Er:YSO2, respectively, and the transitions are labeled according to their site and magnetic subclass. The effect of the magnetic anisotropy is cleary visible; with increasing magnetic field, i.e. decreasing DC g-factor, the collective coupling strength increases. The positions of the resonators below the crystal could be partially identified by laser scanning microscopy (LSM) in a separate experimental setup at 4 K [ZAU06; Zhu+12], see Ref. [Pro+13] for an image of the scan. The evaluation of Er:YSO1 has been discussed in Fig. 5.5. Here, we will focus on the Er:YSO2 sample, which shows a pronounced mode splitting at 250.6 mT indicating strong coupling of resonator #5 to the S_{2a} transition. In the field region between 210 and 265 mT, a faint and narrow absorption line is visible belonging to the same transition. The crossing of this line with the resonances #6-8 effects a complex hybridization pattern due to inductive coupling of the resonators mediated by the spin ensemble coupled to resonator #5. In addition, small dispersive shifts in the field region starting from 200 mT indicate coupling to hyperfine transitions. Figure 5.7 presents a detailed investigation of the avoided level crossing at 250.6 mT. The spectrum in Fig. 5.7(a) shows the avoided level crossing, the absorption dip and two further uncoupled resonances.

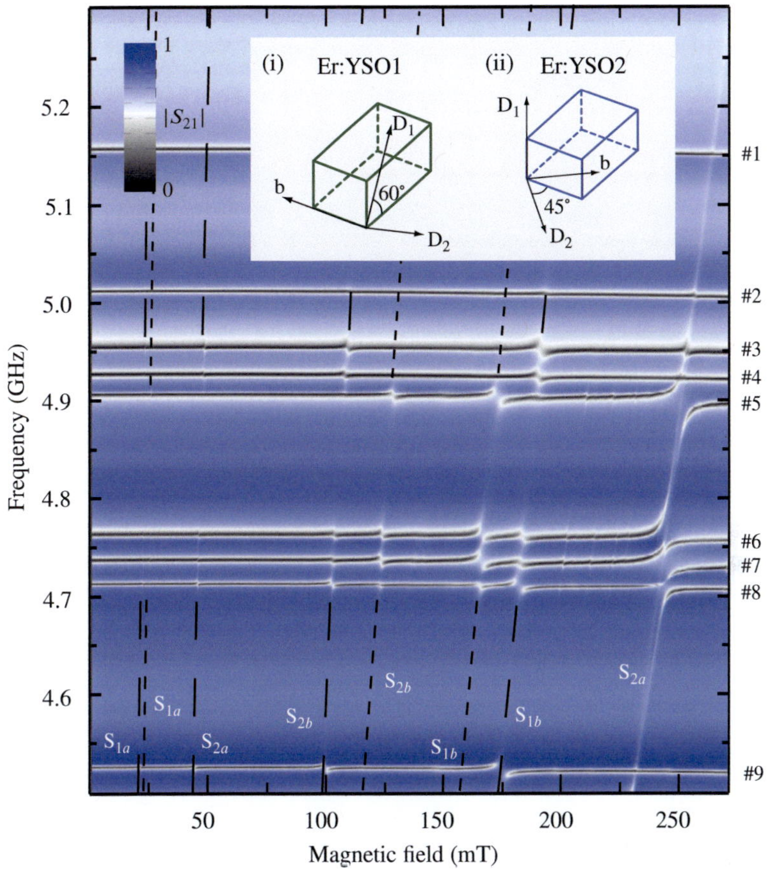

Figure 5.6: On-chip ESR amplitude spectrum of samples Er:YSO1 (inset (i)) and Er:YSO2 (inset (ii)) coupled to an array of nine SC LE resonators. The four electronic transitions of Er:YSO1 are indicated by dashed lines whereas the dotted lines denote the transitions of Er:YSO2.

In oder to analyze the spectrum, the raw power spectrum $|S_{21}(\omega)|^2$ directly in resonance at 253.1 mT is shown in Fig. 5.7(b) and compared to the off-resonant case at 204.3 mT. The full power spectrum in resonance can be described by

$$|S_{21}(\omega)|^2 = B(\omega)\left[1 + \sum_{i=1}^{5} \frac{a_{1i} + a_{2i}(\omega - \omega_i)}{(\omega - \omega_i)^2 + \gamma_i^2/4}\right] . \qquad (5.4)$$

Here, $B(\omega)$ is a second order polynomial accounting for the baseline, ω_i and γ_i denote the frequencies and linewidths of the resonances, respectively. The coefficients a_{1i} and a_{2i} denote the absorptive and the dispersive parts of the resonances. Figure 5.7(b) depicts the fit of Eq. 5.4 to the resonant power spectrum. The resonances are marked in the plot and the dips labeled with $|+\rangle$ and $|-\rangle$ belong to the normal mode splitting. To extract the mode splitting, the raw data is normalized to the baseline and the uncoupled resonances #3 and #4 as well as the absorption dip are subtracted.

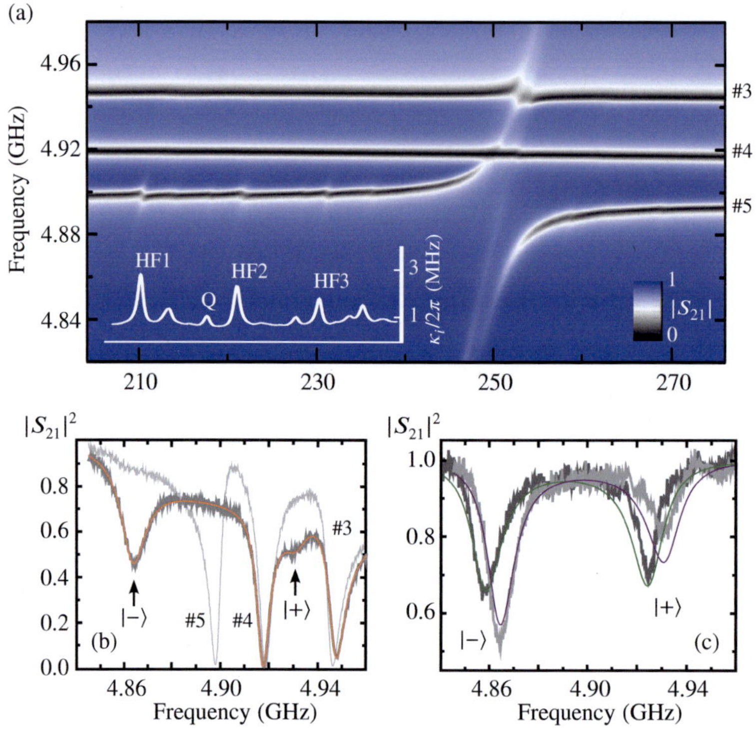

Figure 5.7: (a) Transmission spectrum of resonator #5 strongly coupled to the electronic spin of the S_{2a} transition of the Er:YSO2 crystal. (Inset) The internal damping rate $k_i/2\pi$ of the resonator clearly resolves the hyperfine spectrum of $^{167}Er^{3+}$. The labels "HF1", "HF2" and "HF3" denote magnetic hyperfine transitions and "Q" marks a quadrupole transition. (b) The dark gray line displays the power spectrum $|S_{21}(\omega)|^2$ at the anticrossing at 251.3 mT, which is fit using Eq. (5.4) (orange line). The gray line is the transmitted power measured away from the avoided level crossing at 204.3 mT. (c) The corrected transmitted power spectra taken at 250.6 mT (dark gray line) and at 251.3 mT (gray line) clearly show a normal mode splitting. The solid green and purple lines show the fit to these splittings, respectively.

Figure 5.7(c) presents the corrected spectra at two different spin-cavity detunings $(\omega_S - \omega_c)/2\pi$ of 14 MHz at 251.3 mT (gray line) and 1 MHz at 250.6 mT (dark gray line). Two clearly separated dips are visible confirming the strong collective coupling of resonator #5 to the S_{2a} sub-ensemble. The normal mode splitting is fitted to the model of two coupled quantum harmonic oscillators, see Sec. 3.2.4. The results are presented in Fig. 5.7(c) by the green and purple curves. The fit has been performed in two steps. First, the external $\kappa_c/2\pi = 4.7$ MHz and internal $\kappa_i/2\pi = 0.7$ MHz decay rates of resonator #5 have been determined far away from the ALC at 204.3 mT. These values were set constant for the subsequent fit of the normal mode splitting. The fit yields a collective coupling strength of $\upsilon_N/2\pi = 34 \pm 0.5$ MHz and a HWHM inhomogeneous spin linewidth of $\Gamma_2^\star/2\pi = 12 \pm 0.25$ MHz, amounting for a cooperativity $C = \upsilon_N^2/\kappa_l\Gamma_2^\star$ of approximately 17.8.

To investigate the influence of the ^{167}Er hyperfine transitions, the decay rate of the resonator has been investigated, which is shown in the inset of Fig. 5.7(c). Due to the strong hyperfine interaction, the hyperfine spectrum is rather complex below 6 GHz and amounts for approximately 14 lines in our magnetic field range (205-280 mT). Therefore, it was impossible to identify all transitions with confidence. Beyond 6 GHz, the hyperfine spectrum is dominated by the Zeeman shift and appears as a regular pattern such that the classification is easier, see Refs. [Bus+11; Gui+06]. In general, two types of transitions are visible in the spectrum, hyperfine transitions with constant nuclear spin projection $\Delta m_I = 0$ (denoted ''HF'' in the inset) and weaker coupled (partially forbidden) quadrupole transitions with $\Delta m_I \pm 1$ (labeled ''Q'' in the inset). All of them are weakly coupled and the evaluation of the resonator's damping rate dependency yields a coupling strength of 4 MHz with a linewidth of 7.6 MHz for transition ''HF1''. For the quadrupole transition ''Q'', a collective coupling strength of 1.8 MHz and a linewidth of 7.1 MHz is determined.

Table 5.1 summarizes the results of the microwave spectroscopy. For both samples, the coupling strength scales inversely proportional to the DC g-factor. This is in agreement with the other orientations of sample Er:YSO1 (see Fig. 5.5(a)) and the theoretical considerations in Sec. 2.3.3. Apart from the last two transitions of Er:YSO2, $\Gamma_2^\star$ scales with the DC g-factor. The increase of the inhomogeneous spin linewidth with increasing magnetic field can be explained by inhomogeneities in the spatial distribution of the Zeeman field. However, transitions with a large g-factor are more susceptible to magnetic field fluctuations and inhomogeneities in the magnetic field, which broadens the inhomogeneous spin linewidth, too. For a well aligned sample, the latter effect is dominant and explains the decrease of the linewidth of the S_{2a} transition of sample Er:YSO2 with DC g-factor of 1.4. It should be noted that an ESR study of this crystal on a X-band Bruker Elexsys 580 spectrometer at a temperature of 6.3 K suggests minimal linewidths of 14 and 25 MHz for the low and high field transitions, respectively.

Crystal	Er:YSO2$_{\#5}$ ($\theta = 48°$)			Er:YSO2$_{\#9}$ ($\theta = 43.5°$)		
transition	g-factor	$\upsilon_N/2\pi$	$\Gamma_2^\star/2\pi$	g-factor	$\upsilon_N/2\pi$	$\Gamma_2^\star/2\pi$
S_{1a}	13.4	4.4	23	15.2	4.1	20
S_{2b}	2.7	13	34	3.3	13	24
S_{1b}	1.6	22	32	1.8	21	26
S_{2a}	1.4	34	12	7.3	8	28

Table 5.1: Measured g-factors, collective coupling strengths υ_N, and linewidths $\Gamma_2^\star$ in MHz of all four Er electronic spin transitions.

The narrow absorption line which traverses the spectrum (see Fig. 5.6) is caused by spins directly absorbing the microwave signal from the transmission line. This effect has also been detected in a highly doped ruby crystals [Sch+10; Cla+13]. However in our experiment, the spin concentration is $n_{\mathrm{Er}} \sim 7 \times 10^{17}\,\mathrm{cm}^{-1}$, which is at least an order of magnitude lower. We estimate the number of absorbing spins to be approximately $N_s \sim 10^{13}$, which have a large AC g-factor of about 15.

We studied the absorption line of the microwave transmission line away from the frequency region of the LE resonator array. At a magnetic field of 273.2 mT, corresponding to a frequency of 5.331 GHz, we measure the microwave absorption spectrum. The excitation power amounts for 0.1 fW and is low enough to avoid any saturation of the spin ensemble. Figure 5.8(a) shows the recorded absorption dip, which has a depth of 12%. A fit to a Lorentzian (solid green line) yields a HWHM linewidth of 13.8 ± 0.4 MHz, which is in good agreement with the linewidth determined by fitting the normal mode splitting of resonator #5 (12 MHz). We note that the here demonstrated direct detection of the spin ensemble may pave the way towards a traveling wave quantum memory, which can store and retrieve photons without the aid of a resonator [Afz+13].

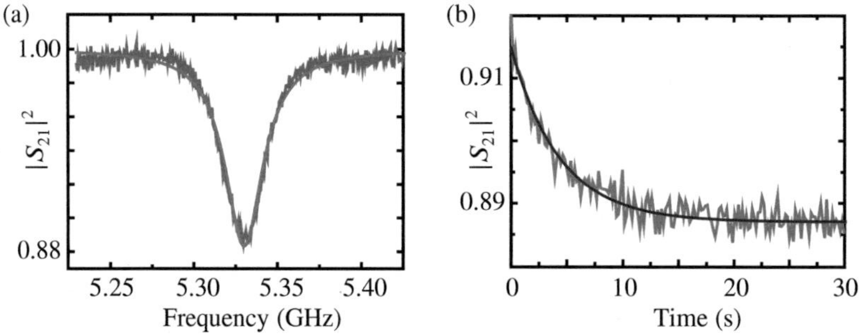

Figure 5.8: (a) Transmission line absorption dip revealing the S_{2a} transition at a magnetic field of 273.2 mT. The green line displays a Lorentzian fit to the spectrum. (b) Reappearance of the absorption signal after a strong saturation pulse. The solid line is an exponential fit to the data.

The presence of an absorption line allows for studying the spin-lattice relaxation dynamics of the spin ensemble directly, see Sec. 2.3.5 for the method. In the experiment displayed in Fig. 5.8(b), a strong and resonant microwave pulse with a power of 0.1 nW and 1 s length saturates the spins which are subsequently probed with a power of 60 fW at the central frequency of 5.330 MHz.

After the strong microwave pulse, the spins relax to their equilibrium level. This is determined by the temperature of the experiment ($T = 20$ mK) yielding $N_2/N_1 = \exp(-\hbar\omega_a/k_B T) \sim 10^{-5}$. Figure 5.8(b) shows the decay of the absorption signal. A fit with an exponential decay curve yields a spin-lattice relaxation time of $T_1 = 4.3 \pm 0.2$ s (solid line). The relaxation time at low temperatures is dominated by a direct process and the measured time is in accordance with the order of magnitude estimation presented in Sec. 2.3.5. A comparison by extrapolation of the relaxation time measured in pulsed ESR [KC80] at 4.2 K gives $T_1 \simeq 3\,s$ in agreement with our value.

Based on these findings, the optimal parameters for coupling LE resonators to Er:YSO are known such that the sample can be studied in further detail. In this experiment, the Er:YSO2 sample was positioned such that it only covers one resonator from the previous study. Note that the coverage of just a single resonator isolates this resonance from the rest of the other resonators due to the strong frequency shift caused by the dielectric constant of the YSO crystal. This allows for a very clean signal. Figures 5.9(a)-(c) present a detailed view on the avoided level crossing of the S_{2a} transition coupled to the LE resonator at different temperatures. While the size of the mode splitting of the electronic transition decreases, some hyperfine transitions appear strongest at 100 mK.

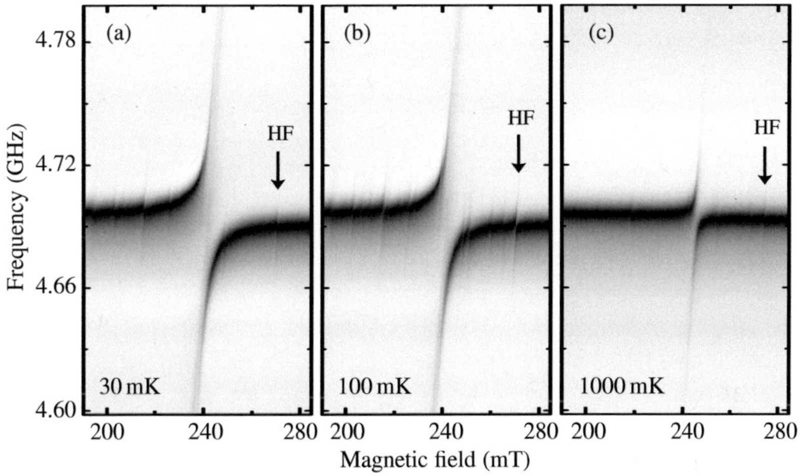

Figure 5.9: Avoided level crossing at different temperatures. Remarkably, the hyperfine transitions are more pronounced at 100 mK, and the system is still strongly coupled at 1 K,.

In order to investigate this effect in greater detail, coupling and linewidth were analyzed as a function of the temperature for both the electronic S_{2a} and the hyperfine transition marked ''HF'' in Fig. 5.9.

Figure 5.10(a) shows the temperature dependence of the collective coupling strength and inhomogeneous spin linewidth. The crossover from the strong to the weak coupling regime is marked by the dotted line. The temperature dependence of the coupling nicely agrees to the theory of paramagnetic crystals as before in the case of weak coupling (gray line), see Fig. 5.3. Due to the pronounced coupling, it is possible to even extract the temperature dependence of one of the hyperfine transitions, which is marked with ''HF'' in Fig. 5.9. Figure 5.10(b) presents the temperature dependence of both collective coupling strength and inhomogeneous spin linewidth. The linewidth increases with increasing temperature from approximately 4 MHz to 5.5 MHz, whereas the coupling strength shows a pronounced maximum at 140 mK. There, the coupling has a value of almost 4.4 MHz that approaches the strong coupling regime because spin and resonator linewidth reside in the same regime. Below 140 mK, the coupling strength decreases quickly to about 2.1 MHz. Towards higher temperatures, the coupling strength decreases to about 2.4 MHz at 1 K.

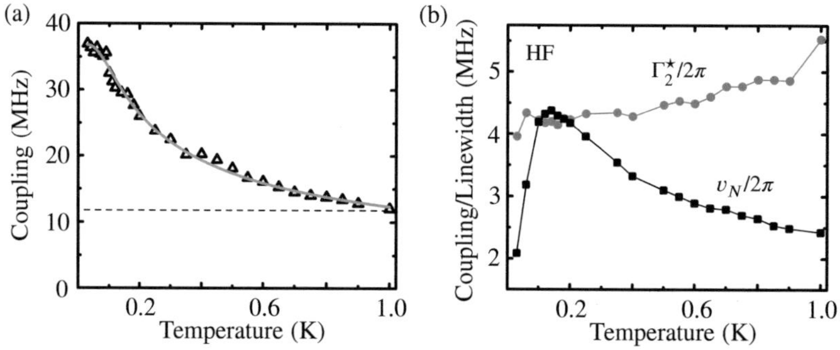

Figure 5.10: (a) Temperature dependence of the collective coupling strength of the S_{2a} transition coupled to the LE resonator. The solid line is a fit to the theory and the dashed line indicates the level of the inhomogeneous spin linewidth. (b) Temperature dependence of coupling strength and linewidth of the hyperfine line "HF". The coupling is strongest around 140 mK and approaching the strong coupling regime.

This behavior can be explained by looking at the level structure of the odd ^{167}Er spin ensemble. The nuclear spin 7/2 couples to the electronic spin 1/2 such that the composite system consists of 16 energy levels. The strong hyperfine interaction effects a zero field splitting of approximately 5.6 GHz for the largest transition. Since our LE resonator has a resonance frequency of 4.7 GHz, we are working in a regime where the Zeeman and the hyperfine interaction are of the same order of magnitude. Consequently, the hyperfine spectrum is rather complex and it is not possible to associate all lines in the spectrum with confidence. Nevertheless, it is possible to draw some qualitative conclusions. At zero temperature, only the lowest $|I,S\rangle = |7/2,-1/2\rangle$ of the 16 levels will be populated. A transition originating from a higher level, e.g. $|3/2,-1/2\rangle \rightarrow |3/2,1/2\rangle$, will possess zero coupling strength because there is no population. As the temperature increases, the lower levels get filled first such that the coupling strength of the transition increases up to the point where the higher states become populated as well. Then, the coupling strength decreases again because it is proportional to the population difference of the corresponding levels $\upsilon \propto \sqrt{N_1 - N_2}$.

Taking the hyperfine tensors provided in Ref. [Gui+06], it is possible to estimate the energy spectrum using the Matlab package EASYSPIN [SS06]. Based on this, we calculate the thermal population of all 16 hyperfine levels as a function of temperature. The occupation probability of a certain level with energy E_i is given by the Boltzmann statistics $p_i = \frac{1}{Z}\exp(-\beta E_i)$, where Z is the partition sum and $\beta = 1/k_B T$. We extract the coupling strengths $\upsilon \propto \sqrt{p_i - p_j}$

of the transitions preserving the nuclear spin momentum[1]. The result is shown in Fig. 5.11. The simulated temperature dependencies show maxima for all but one transition and are in qualitative agreement with the measured dependencies. In principle, these transitions could serve as a sensitive sensor to determine the temperature of the spin system.

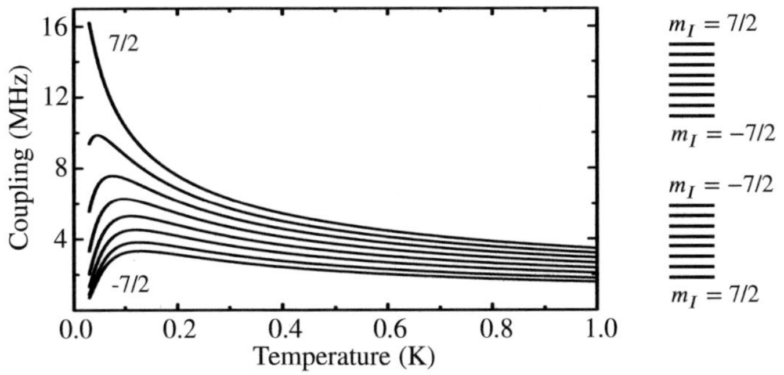

Figure 5.11: Simulated temperature dependencies for the coupling strengths of the hyperfine transitions of ^{167}Er.

In conclusion, we have demonstrated coherent strong coupling of a SC LE resonator to an Er^{3+} spin ensemble. The collective coupling strength shows a paramagnetic temperature dependence in accordance with theory. This is also confirmed for one of the ^{167}Er hyperfine transitions. The spin relaxation time of 4.3 s measured via direct microwave absorption is comparable to the estimations presented in Sec. 2.3.5. From the exploration of the magnetic anisotropy, we can conclude that strong coupling is only achieved at high magnetic fields, i.e., small DC g-factors. In contrast, the low field transitions show weak coupling and a large linewidth.

From these investigations, we have acquired a good understanding of the physics of Er:YSO spin ensembles coupled to LE resonators. However, with regard to applications in SC quantum information circuits, it would be desirable to operate at much lower magnetic fields, which have to be compatible with the Josephson junctions of SC qubits. It is therefore interesting to investigate rare earth ions doped into other host crystals such as $YAlO_3$ with higher symmetry and less anisotropic g-tensors.

5.2 Rare earth ions doped into $YAlO_3$ host crystals

In the Sec. 5.1 it was shown that reaching the strong coupling regime requires trading the coupling strength against the high spin tuning rate originating from large g-factors. This requires large magnetic fields beyond 200 mT to achieve the necessary Zeeman splitting. Since the final goal of these investigations is the integration of rare earth spins into the state

[1] Note that this is a simplified approach, which neglects the strong mixing of the spin levels, which can lead to different allowed transitions.

of the art circuit QED architecture, operating at lower magnetic fields $\sim 10\,\mathrm{mT}$ is desirable [Sen+08]. At larger magnetic fields, the critical currents of the Josephson junctions will degrade significantly. In addition, magnetic vortices in the superconductor increase the losses and limit the overall coherence of the circuit. The magnetic anisotropy of RE doped crystals strongly depends on the properties of the host material, e.g. the symmetry of the crystal. Choosing a proper host material can overcome the limitations of Er:YSO spin ensembles and enable strong coupling at much lower magnetic fields. Here, a $YAlO_3$ (YAlO) crystal is chosen as a host for Nd^{3+} and Er^{3+} ions [ARM97; AR02]. Both Nd and Er are interesting because they both exhibit optical transitions inside standard telecommunication bands.

$YAlO_3$ belongs to the orthorhombic space group D_{2h} and its unit cell is subdivided into four distorted perovskite cells, which create four magnetic subclasses for the Y^{3+} ions [ARM97]. The mirror symmetry of the Y^{3+} ions in the (001) plane and the inversion symmetry through the aluminum sites reduces the number of effective subclasses to two. So, in contrast to YSO, one expects to see at maximum two sub-ensembles in the ESR spectrum. The hyperfine spectrum of ^{167}Er doped YAlO differs from Er:YSO and exhibits a zero field splitting with a maximum transition frequency of almost 4 GHz, see Ref. [Tka+14] for a simulation of the spectrum. In the case of Nd:YAlO, the maximum zero field splitting (ZFS) amounts for approximately 3.5 GHz. The frequency range of the ZFS would allow one to directly couple a flux qubit to the spin ensemble at zero magnetic field [Mar+10; Zhu+11]. Moreover, the hyperfine spectrum features so-called zero first-order Zeeman transitions, which can enhance the coherence properties of the spin ensemble because of their insensitivity to magnetic field fluctuations to first order [McA+12]. According to Ref. [AR02], the principal values of the g-tensor of Er:YAlO are $g_x = 8.98$, $g_y = 8.13$ and $g_z = 2.73$. For Nd:YAlO, $g_x = 2.83$, $g_y = 2.58$ and $g_z = 1.69$ have been determined.

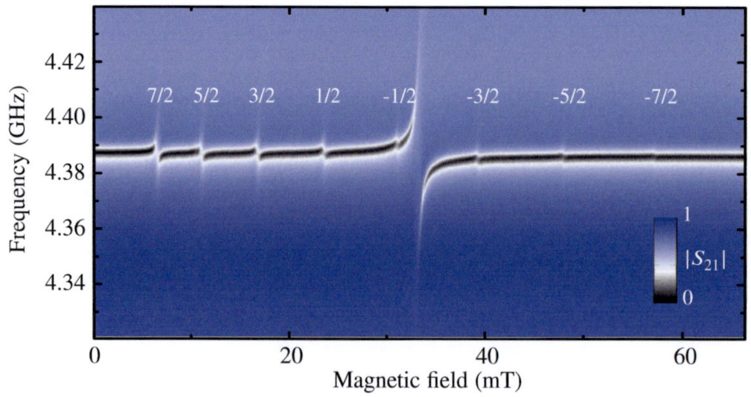

Figure 5.12: On-chip ESR spectrum of Er:YALO coupled to LE resonator #1. The numbers give the nuclear spin numbers for the eight hyperfine transitions.

The Er:YALO sample being investigated in the following is a 50 ppm erbium doped crystal from Scientific Materials Inc. The experimental setup is comparable to the previous chapter apart from two differences. The employed SC chip is a 3 LE resonator chip and the crystal is fixed by gently pressing it with a teflon screw onto the SC chip covering two LE resonators. In this configuration, the resonators operate in the frequency band around 4.4 GHz and exhibit loaded and internal quality factors of $Q_l \sim 660$ and $Q_i \sim 2500$. The magnetic field is applied in the (001) plane of the crystal and forms an angle of 10° with the crystallographic *b* axis.

Figure 5.12 shows the on-chip ESR spectrum of LE resonator #1 coupled to the Er:YAlO spin ensemble. Only one electronic transition with a DC g-factor of 9.4 is detected. It shows a pronounced avoided level crossing. The collective coupling strength is $\upsilon_N = 32.2$ MHz and the inhomogeneous spin linewidth is measured to be $\Gamma_s^\star = 16$ MHz. In comparison to our experiments with Er:YSO, both coupling and linewidth are comparable but the strong coupling is already achieved at 35 mT, which is a factor of seven improvement over the required field of 250 mT for Er:YSO. In this experiment, the two magnetic classes overlap. We believe that a small misalignment of the magnetic field remains, which gives rise to an increased inhomogeneous spin linewidth such that an improved alignment could reduce $\Gamma_2^\star$ further.

The spectrum presented in Fig. 5.12 also shows the hyperfine transitions of the odd ^{167}Er isotope. In contrast to ^{167}Er:YSO, we are operating in a frequency regime above the ZFS. Therefore, all hyperfine lines are visible and can be associated to the respective nuclear momenta −7/2 to 7/2. The following table lists the coupling strengths and linewidths of the transitions with nuclear spin momentum m_I.

m_I	7/2	5/2	3/2	1/2	−1/2	−3/2	−5/2	−7/2
$\upsilon_N/2\pi$ (MHz)	9.7	7.3	7.6	7.2	3.8	5.7	5.7	5.0
$\Gamma_2^\star/2\pi$ (MHz)	31	27	29	26	15	22	33	25

From lower to larger fields, the coupling strength of the transitions decreases, which can be attributed to the thermal population of the hyperfine state. Since we are only 400 MHz away from the ZFS, the transition matrix elements of the hyperfine transitions are not unity but rather vary in a complex way. This prevents a more detailed study of the thermal population of the levels and their coupling strength. However, by repeating the same experiment with a resonator operating at higher frequencies (e.g. 9 GHz) one can circumvent the problem and even use the hyperfine spectrum to measure the temperature of the spin system by comparing the individual coupling strengths. A complete characterization of the Er:YAlO spin ensemble is given in Ref. [Tka+14]. For the purpose of this thesis it is sufficient to note that strong coupling can be achieved at low magnetic fields of 35 mT. Even though such a field is not yet suited for the direct integration with SC qubits, it is much easier to deal with an can be shielded from the rest of the circuitry. Our investigation showed a pronounced hyperfine coupling where the low field transitions appear strongest. By choosing an isotopically purified ^{167}Er:YAlO spin ensemble, one may be able to operate even at zero field. The data suggests for a purified sample an approximate collective coupling strength to the LE resonator of ~ 22 MHz.

A feasible quantum computer technology requires the integration of local memory elements directly on the SC chip. The local integration of spins via ion implantation is presented in Sec. 5.3. Here, the performance and coupling of SC LE resonators are investigated, which were directly fabricated on a 1000 ppm Nd doped $YAlO_3$ substrate from the MTI company. These substrates had never been used before for the fabrication of SC circuits and required an adoption of the fabrication process by our collaborator S. Wünsch. Moreover, the dielectric permittivity of $YAlO_3$ was only given in a range between 16 and 20, which made it impossible to achieve a good impedance matching to 50 Ω. The microwave spectroscopy is carried out in the usual way and the crystal has a similar orientation as the Er:YAlO sample discussed before such that the DC g-factor is optimized.

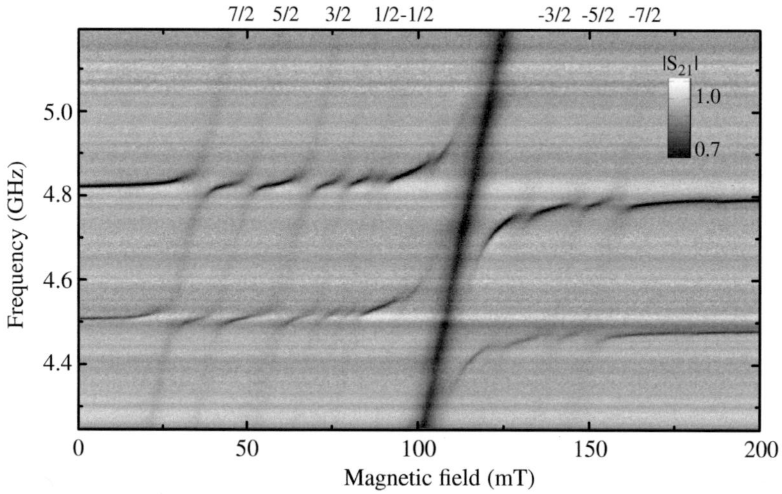

Figure 5.13: Avoided level crossings between the two LE resonators and the Nd spin ensemble. The coupling to the hyperfine transition is clearly visible.

Despite the expected impedance mismatch, the sample shows a strong signal. Due to the impedance mismatch, the baseline of the spectrum shows low frequency modulations. These can be clearly distinguished from the pronounced and narrow signal of the LE resonators. Figure 5.13 presents the baseline corrected spectrum. The spectrum shows several pronounced dispersive shifts and a large avoided level crossing (ALC) at a magnetic field of approximately 110 mT. This is attributed of the electronic transition of the even Nd isotopes. There are two odd isotopes (^{143}Nd and ^{145}Nd) with a nuclear spin of 7/2 but different nuclear magnetic moments. ^{143}Nd with 12.2% abundance contributes more to the signal compared to ^{145}Nd with 8.3% natural abundance [HL11]. The respective hyperfine tensors are slightly different and, to our knowledge, only the hyperfine tensor of ^{143}Nd has been measured so far [AR02].

Another prominent feature of the spectrum are the intense absorption lines traversing the spectrum. In order to extract the coupling strength and linewidth of all transitions, it is necessary to apply a common fit, including the electronic absorption line, because they are partially overlapping and influence each other. Therefore, a two dimensional fit on the entire spectrum was performed. The DC g-factor of the electronic transition is determined to be 2.97 and the collective coupling strength yields $\upsilon_N/2\pi = 199\,\text{MHz}$ with an inhomogeneous spin linewidth of $\Gamma_2^\star/2\pi = 21\,\text{MHz}$. Thus, we operate in the strong coupling regime, and it is the strongest coherent coupling ever measured with paramagnetic spin ensembles in a hybrid circuit QED architecture. The collective coupling strength of the first hyperfine transition amounts for $\upsilon_N/2\pi = 97\,\text{MHz}$ but has a linewidth of $\Gamma_2^\star/2\pi = 144\,\text{MHz}$. The large linewidth originates from the fact that both odd Nd isotopes contribute to the coupling. However, due to their different hyperfine interactions their frequencies deviate from each other causing this large inhomogeneous spin linewidth. Due to the visibility of the absorption lines of both the electronic and the hyperfine transitions, it is possible to study the spin Hamiltonian over a much broader frequency range.

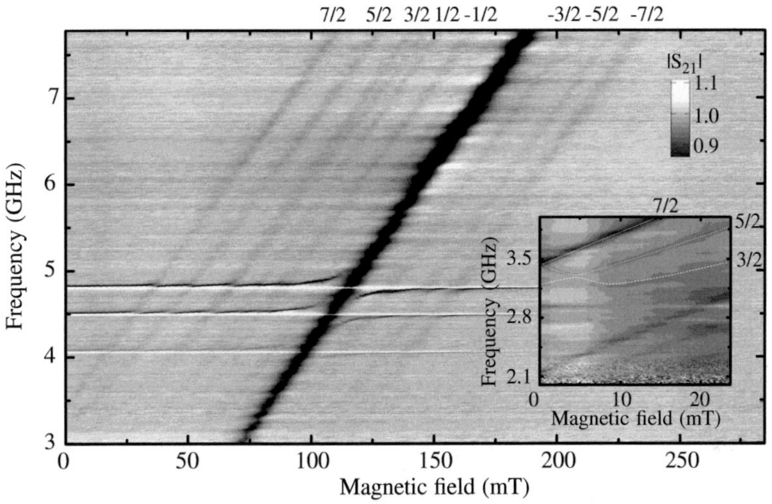

Figure 5.14: Absorption lines due to direct MW-ensemble coupling. The strong line is the electronic transition of the even Nd isotopes. The fainter lines belong to the two odd isotopes ^{143}Nd and ^{145}Nd with 12.2% and 8.3% natural abundance, respectively [HL11]. (Inset) Absorption at low magnetic fields. The 7/2 line is clearly visible down to zero field. $|S_{21}|$ values larger than unity are an artifact of the normalization procedure, which divides all the data by the frequency spectrum at maximum field in order to remove the oscillations of the baseline.

Figure 5.14 presents the absorption line spectrum from 3 GHz to 7.7 GHz. The observable frequency range is only limited by the operational band of the HEMT amplifier and the employed cryogenic microwave isolator (both 4-8 GHz). The linewidth of the absorption line of the electronic spin transition shows a clear magnetic field dependence. Its depth in the power spectrum increases from approximately 35% at 3.48 GHz to 65% at 6.87 GHz. The absorption dip is fitted with a Lorentzian and yields a linewidth of $\Gamma_2^\star/2\pi = 91 \pm 5\,\text{MHz}$ at 3.48 GHz (83.1 mT) and 114 MHz at 6.87 GHz (166.2 mT). This is larger than the linewidth obtained by fitting the avoided level crossing. The larger inhomogeneity may originate from magnetic field inhomogeneities along the 1 cm long transmission line which is much more susceptible to those compared to the smaller dimensions of the LE resonators of $0.6\times0.6\,\mathrm{mm}^2$. It should be noted that despite the large linewidth determined by direct fitting of the absorption dips, this linewidth would still suffice to enter the strong coupling regime due to the large collective coupling strength of the LE resonator to the ensemble of $\upsilon_N/2\pi = 199\,\mathrm{MHz}$. Down to about 4 GHz, the hyperfine absorption lines are clearly visible. Some absorption lines split at larger magnetic fields indicating the presence of the two odd Nd isotopes. At lower frequencies, the hyperfine interaction dominates the Zeeman interaction, leading to a strong mixing of the eigenstates in the uncoupled basis $|S, I\rangle$ and reducing the transition matrix elements. Specifically, below 4 GHz, the hyperfine absorption dips become fainter which is accompanied by a bending of the absorption lines towards the vertical axis of the plot. The inset of Fig. 5.14 presents a fine scan around zero magnetic field. Although we are operating below the optimal band of the amplifier, it is possible to identify the 7/2, 5/2 and 3/2 states and follow them back to their zero field splitting. At zero field, the 7/2 state shows the strongest signal and possesses a transition frequency of approximately 3.45 GHz. The reason for the weaker absorption signal at lower fields is mainly caused by a reduction of the transition dipole moment, which is confirmed by simulating the hyperfine spectrum using the data from Ref. [AR02]. In contrast to the simulation, which predicts a vanishing transition matrix element at zero field, we can still identify the 7/2 absorption line. Therefore, it is possible to couple a resonator at zero magnetic field to the 7/2 transition while remaining compatible with SC Josephson qubits. This could be realized by an isotopically purified ^{143}Nd:YAlO spin ensemble. Furthermore, both the 5/2 and 3/2 absorption lines show a local minimum and maximum at 4 mT, respectively. Those present so-called zero first order transitions, which prolong the coherence time of the spin system due to their first order insensitivity to magnetic field fluctuations.

5.3 Focused ion beam implanted Er^{3+} in Y_2SiO_5

This section focuses on a further aspect towards the practical implementation of erbium based hybrid systems. In a SC quantum processor circuit, memory elements are required to be placed at specific positions, where they can fulfill their tasks without interfering with the rest of the circuitry. One possible approach is to locally implant spins into an empty crystal or directly into the substrate on which the circuit is fabricated [Kol+12]. In our work, we use a focused ion beam (FIB) to implant Er^{3+} ions into an undoped Y_2SiO_5 (YSO) crystal with high spatial resolution[Kuk+14]. We then perform circuit QED experiments on these crystals and confirm the successful implantation of erbium ions by studying the electron spin resonance properties at the single photon level. The results of this work have been published in Ref. [Pro+14a]. Recently, a related work by Wisby et al. reports weak coupling ($\sim 1\,\mathrm{MHz}$) of SC resonators to an implanted Gd^{3+}:Al_2O_3 spin ensemble [Wis+14]. However, in our work, Y_2SiO_5 is used as a host material for the Er^{3+} ions, which promises long optical and spin coherence times [Böt+06b; Ber+07; Rak+09]. In addition, we achieve collective coupling strengths which are an order of magnitude larger compared to the work by Wisby et al. Figure 5.15(a) shows a schematics of the ion implantation, which has been performed at RUB Bochum by our collaborators N. Kukharchyk and A. D. Wieck [Kuk14]. The crystal's surface is divided into several areas where we implant Er^{3+} ions in order to study the effect of the amount of implanted ions per area (ion fluence) on the coupling strength and inhomogeneous spin linewidth. The Erbium ions were implanted into the YSO crystals in an EIKO-100 FIB system with an energy of 300 keV. The ions were extracted from an $Au_{78.4}$ Er_{10} $Si_{11.6}$ liquid metal ion source (LMIS), developed by A. Melnikov at al. [Mel+02]. The ion beam was accelerated to an energy of 300 keV and separated into its ion isotopes by a built-in Wien filter. The resolution of the filter does not allow for a fine separation between the Erbium isotopes, but it was possible to minimize the amount of the unwanted ^{167}Er isotope to less then 5%. For comparison, the natural abundance amounts for approximately 20%. The amount of ion fluence F is estimated by $F = I\frac{t}{S}$, where I is the ion current, t is the dwell-time of the beam on a single drawing point, and S is the area covered by the ion beam.

The penetrating ions trigger a number of processes in the crystal, mainly ion-ion collisions resulting in lattice defects. This implantation process has been simulated by our collaborators employing the SRIM software [ZZB10], which yields an ion distribution with a mean depth of 60 nm and 39 nm mean deviation [Kuk+14]. In order to repair the damage of the crystal lattice caused by the ion beam, thermal annealing is required. At RUB Bochum, both rapid thermal annealing (RTA) and 1.5-2 hours long-time annealing have been performed.

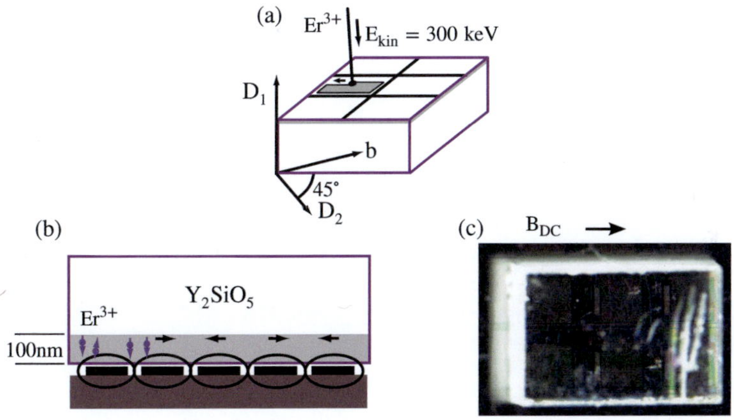

Figure 5.15: (a) Sketch of the implantation process. A focused Er^{3+} ion beam irradiates several areas of an undoped Y_2SiO_5 crystal. (b) Schematics of the coupling geometry. (c) Picture of the implanted YSO crystal on top of a superconducting chip with an array of nine LE resonators. Each implanted area fits to the size of one LE resonator.

There are two ways to investigate the properties of implanted ions, either optically, i.e. by confocal photoluminscence [Kuk+14], or in the microwave frequency range using electron spin resonance (ESR). In the case of a magnetic ion, ESR allows one to obtain information about the properties of spins in solids, such as g-factors, relaxation and coherence times, and inhomogeneous linewidths [SE01]. Typically, 3D resonators deployed in ESR measurements probe the spins distributed over the whole crystal volume. In contrast, on-chip ESR allows to study the surface region of the paramagnetic sample [Sch+10; Bus+11; Kub+12; Sig+14]. Since the implanted ions only occupy a region 100 nm below the surface, on-chip ESR is a convenient tool for this investigation, see Fig. 5.15(b).

Figure 5.15(c) shows a picture of the YSO crystal magnetically coupled to an array of superconducting lumped element (LE) resonators similar to the experiments presented in Sec. 5.1. In our work, we investigate two samples (YSO2 and YSO4), which have 5 and 4 implanted areas, respectively. Each implanted area covers one specific LE resonator. The LE resonators covered by the Er:YSO crystal possess loaded quality factors of $Q_l \sim 650$. The resonance frequencies of the LE resonators occupy the frequency range from 4 to 5 GHz.

Figure 5.15(b) presents a schematic cross-section of the coupling area. The oscillating magnetic field of the microwave penetrates several microns into the crystal, see Ref. [Pro+13] for a simulation of the AC field. A DC magnetic field is applied along the surface of the SC chip.

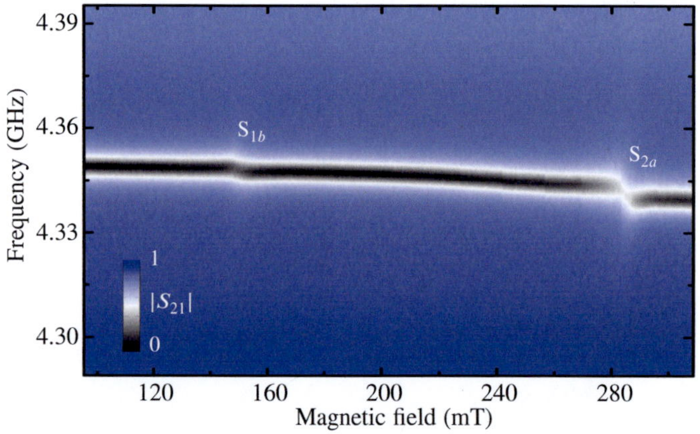

Figure 5.16: ESR transmission spectrum of the YSO4 crystal, area #3 coupled to one of the LE resonators. Two dispersive shifts of the resonator are visible in the spectrum due to the magnetic coupling to the electronic spin transitions S_{1b} and S_{2a} of the erbium ions

Figure 5.16 shows the on-chip ESR spectrum of the YSO4 sample area #3. The plot displays the transmitted amplitude $|S_{21}|$ in color as a function of the magnetic field and the probe frequency. The resonator shows up as a black line which is distorted by two dispersive shifts. There, the resonator couples weakly to the Er spins [Sch+10; Bus+11]. From the spectrum, we extract the collective coupling strength υ_N and inhomogeneous spin linewidth $\Gamma_2^\star$ by fitting the weak coupling theory described in Sec. 3.2.3 to the decay rate of the resonator. In the following, we will focus on the high field transition of our samples, which possess DC g-factors of 1.09 (YSO4) and 1.03 (YSO2), respectively. The orientation of the crystals is shown in Fig. 5.15(a).

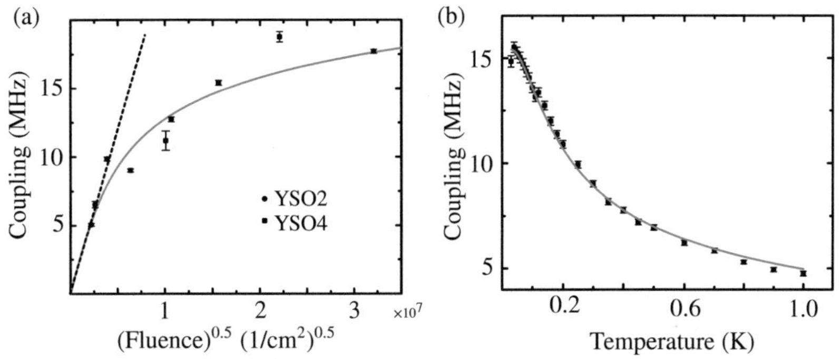

Figure 5.17: (a) Collective coupling vs. the square root of the fluence. The dashed straight line shows the expectation for a 100% effective implantation. The gray line is a fit to an empirical theory, see text for details. (b) Temperature dependence of the collective coupling strength of sample YSO4 area #3. The temperature dependence follows the expectation for a paramagnetic crystal (gray line).

The on-chip ESR spectra of all samples are analyzed by comparing the collective coupling strength υ_N and inhomogeneous spin linewidth $\Gamma_2^\star$ to the implanted fluence. Figure 5.17(a) displays the extracted coupling strengths versus the square root of the fluences. If the implantation process had the same efficiency for all fluences, one would expect the data points to follow a straight line because $\upsilon_N \propto \sqrt{F}$. However, the coupling strength is not proportional to the square root of the incident fluence F and the deviation increases for larger fluences.

In order to interpret this result, it is important to emphasize that the entire sample preparation process consists of FIB irradiation and subsequent annealing in Ar atmosphere. Both processes can provide a contribution to the reduced implantation efficiency. Two samples, which were only treated with RTA, show no ESR response. We found that long term annealing ($\sim$ 1.5-2 hours) is crucial in order to detect the erbium ions with on-chip ESR. However, a large inhomogeneous spin linewidth remains and does not show a significant dependence on the fluence.

There are two major contributions to the inhomogeneous spin linewidth: Dipole-dipole interaction and local distortions of the crystal field which modulate the effective g-factor. Using Refs. [Bro+11; BC08], we can estimate an upper limit for the contribution to the inhomogeneous broadening due to dipole-dipole interaction. Assuming the ion distribution calculated by our Monte Carlo simulation, $\Gamma_2^\star/2\pi$ varies from tens of kHz for the lowest fluence to approximately 80 MHz for the sample with the largest fluence. Since the linewidth always resides at a high level, we conclude that the distorted crystal field dominates the linewidth. The details of the implantation and annealing processes are complex and a more detailed study would be needed to model the entire process.

Since we do not know exactly the distribution of the ions in the crystal and the distortion of the lattice, we use the known implantation process parameter fluence to motivate a simplified empirical model. From all erbium ions per area F penetrating the sample, only a subset

$D \leq F$ will finally be a in an appropriate configuration. Therefore, merely these ions contribute to the ESR signal and $\upsilon_N \propto \sqrt{D}$. We assume that the implantation yield (including post-annealing) is exponentially reduced by the amount of ions per implanted area D (dose), $\dot{D} = k\ \exp(-D/D_c)$. Here, $k = I/S$ is the intensity of the FIB and D_c denotes a critical dose where local interaction effects in the crystal start to dominate over the enhanced amount of erbium ions. The total number of incident ions per area is given by the fluence $F = kt$, where t denotes the total implantation time for a given area S. The effective implanted dose D with respect to the fluence is given by

$$D(F) = \ln\left(1 + \frac{F}{D_c}\right) D_c\,. \tag{5.5}$$

Accordingly, the collective coupling scales as $\upsilon_N \propto \sqrt{D(F)}$. The fit to the experimental points is presented in Fig. 5.17 and yields $D_c \approx 10^{13}\,\mathrm{cm}^{-2}$. Only in the limit of low fluence $F \ll D_c$, we observe that the collective coupling is proportional to the total number of implanted ions.

In order to check whether the implanted ions form a paramagnetic spin ensemble, we studied the temperature dependence for sample YSO4 area #3. Figure 5.17 shows the measured temperature dependence of the collective coupling strength υ_N, which nicely follows the theory of paramagnetic crystals

$$\upsilon(T) = \upsilon_0 \sqrt{\tanh\left(\frac{\hbar\omega}{2k_B T}\right)}\,. \tag{5.6}$$

Thus, the implanted ions can be modeled as a system of the independent spins [Bus+11]. In contrast to the coupling strength, the inhomogeneous spin linewidth $\Gamma_2^\star/2\pi$ is about 70 MHz and stays constant in the temperature range from 30 mK up to 1 K. Since crystals grown by the Czochralski method are known to have a much smaller linewidth at low temperatures (12 MHz for 200 ppm doping concentration, see Sec. 5.1), we assume that the distorted crystal field dominates the inhomogeneous broadening over dynamic effects in this temperature range.

In the limit of weak coupling where $\upsilon_N \ll \Gamma_2^\star$ and the cooperativity $C = \upsilon_N^2/\kappa\Gamma_2^\star < 1$, we can study the spin-lattice relaxation dynamics directly. Here, we employ a technique similar to Sec. 5.1 where our weakly coupled cavity is equivalent to a long transmission line. The details of the method are given in Sec. 3.3. An initial strong microwave pulse saturates the spin ensemble and the depth of the resonator dip increases significantly. The resonator is then continuously probed at low power in order to observe the decrease of the resonator dip while the spin ensemble is relaxing back to its equilibrium position. We observe a double exponential decay with two time scales $T_{\mathrm{SLR}}^a \approx 1\,\mathrm{s}$ and $T_{\mathrm{SLR}}^b \approx 10\,\mathrm{s}$. This suggests that the energy relaxation is dominated by a direct process on short timescales plus an indirect process on larger timescales. Typically, the spin-lattice relaxation time T_1 at low temperatures ($\hbar\omega \gg k_B T$) is dominated by a direct process $1/T_1 \propto (\hbar\omega)^5 \coth(\hbar\omega/2k_B T)$, see Sec. 2.3.5.

Table 5.2 summarizes the results of our on-chip ESR investigations of implanted Er:YSO samples and compares them to a doped as-grown Er:YSO crystal. We note here, that the implanted crystals have a similar orientation as the bulk doped crystal investigated in Sec. 5.1. The effective local concentration and number of spins are calculated from F and using Eq. (5.5). All implanted samples couple weakly to the LE resonators because $\upsilon \ll \Gamma_2^\star$. The linewidth $\Gamma_2^\star$ does not show a significant dependence on the coupling strength or fluence. The value of the calculated average single-spin coupling is in accordance with our estimation of $\bar{\upsilon}_1/2\pi \sim 70\,\mathrm{Hz}$ using the inductance of the LE resonator [Wue+11] and the simulated magnetic field [Pro+13] with an AC g-factor of 15, see also Sec. 3.2.1. We detect no significant difference between the YSO2 and the YSO4 samples.

To summarize this section, erbium ions were implanted by focused ion beam irradiation of a Y_2SiO_5 substrate. The ions possessed an energy of 300 keV and penetrated up to 100 nm inside the substrate. A subsequent annealing of the samples in argon atmosphere at 1200°C for 1.5-2 hours turned out to be crucial. The implanted spin ensemble was characterized by on-chip ESR spectroscopy at 20 mK and at the single microwave photon limit. The collective coupling strengths of the implanted spins vary with the implantation fluence from 5 to 18.7 MHz, and exceed the typical dissipation rates of SC circuits. The temperature dependence of the coupling strength shows paramagnetic behavior and the inhomogeneous spin linewidth is 5 to 10 times larger compared to a bulk doped Er:YSO crystal grown by the Czochralski method. In order to reduce the spin linewidth, further investigations of the annealing process are required. We believe that implantation into heated substrates will be crucial for an optimal process. This work paves the way towards the local integration of erbium spins in SC quantum circuits. This concept can be developed further to enable reversible conversion of microwave and telecom C-band photons by combining microwave and optical waveguides on the same crystal [SPT13; OBr+14].

crystal ID	YSO2					YSO4				bulk
implanted area	#1	#2	#3	#4	#5	#1	#2	#3	#4	Er:YSO
$\upsilon_N/2\pi$ (MHz)	5.06	6.53	9.85	12.73	17.71	9.01	11.18	15.41	18.77	34
$\Gamma_2^\star/2\pi$ (MHz)	68.1	107	105	105	112	48.6	136	69.2	93.1	12
F (10^{12}cm^{-2})	4.80	6.73	15.0	114	1030	40.6	102	244	487	–
$\tilde{c}_{\text{eff}}{}^{\ddagger}$ (ppm)	35	46	81	225	414	145	215	288	348	200
$\tilde{N}_{\text{eff}}$ ($\times 10^9$)	14.1	18.5	32.9	90.8	167	58.4	86.8	116	141	$\sim 10^3$
$\bar{\upsilon}_1/2\pi$ (Hz)	85	96	109	85	87	75	76	90	100	76
T^a_{SLR} (s)	†	†	1.13	1.10	1.19	1.83	2.91	1.9	(2)*	4.3
T^b_{SLR} (s)	†	†	7.36	8.16	7.49	12.2	15.6	11.9	(14)♯	–
annealing param.	2 h at 1200°C in Ar atmosphere					1.5 h at 1200°C in Ar atmosphere				–

‡assuming an average 60 nm thick layer, *uncertainty ≈ 30%, ♯uncertainty ≈ 25%, †SLR signal too weak

Table 5.2: Comparison of coupling strength υ, inhomogeneous spin linewidth $\Gamma_2^\star$, fluence F, effective local concentration $\tilde{c}_{\text{eff}}$, effective number of spins $\tilde{N}_{\text{eff}}$ coupled to the transition, the average single spin coupling strength $\bar{\upsilon}_1$ and spin lattice relaxation times T^a_{SLR}, T^b_{SLR} of all samples with implanted Er^{3+} ions and a doped as-grown crystal.

5.4 3D cavity QED experiments with $Er^{3+}:Y_2SiO_5$

In Sec. 5.1, the DC g-factor and the collective coupling anisotropy has been studied. Already at this stage, it was pointed out that the oscillating magnetic field of the LE resonator is largely inhomogeneous in strength and direction. Therefore, the collective coupling strength is given by an integral over different AC g-factor contributions. This two-fold inhomogeneity has challenging implications regarding the implementation of a quantum memory. The spatial and g-factor inhomogeneities lead to an inhomogeneous distribution of the Rabi frequencies of the spins. Note that these issues can be addressed by optimal control pulse schemes [Sig+14], which becomes relevant for the efficient implementation of a multimode quantum memory, see also Sec. 6.5.

There is a way to overcome the problem of the AC field inhomogeneity by using a three dimensional geometry for the microwave resonator. In collaboration with Andrej Tkalčec [Tka13] and the group of Prof. M. Tobar, a cylindrical microwave cavity has been developed. Figure 5.18(a) shows a photo of the 3D copper cavity. In oder to reduce the mode volume, it has been loaded with a sapphire cylinder confining the electrical field. The 50 ppm Er^{3+} doped YSO crystal is placed in the center of the cylinder supported by two teflon corks. Figure 5.18(b) presents a schematics of the coupling geometry. The so-called TE_{011} mode ensures a homogeneous magnetic AC field at the position of the crystal (dashed line), which is also confirmed by simulations, see Ref. [Pro+14c].

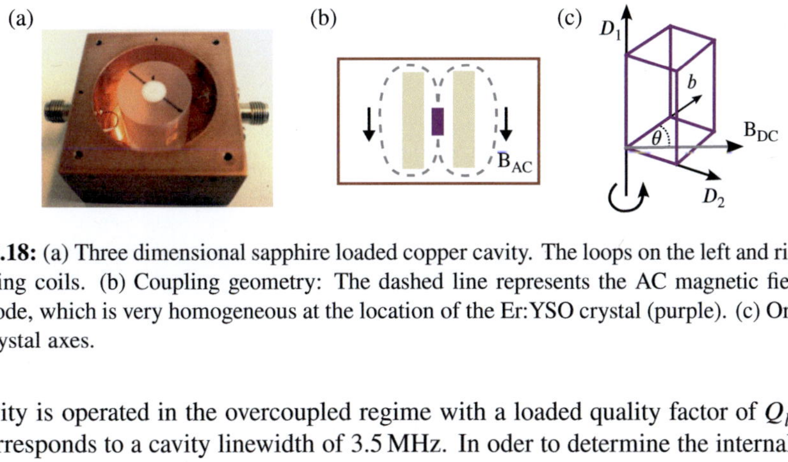

Figure 5.18: (a) Three dimensional sapphire loaded copper cavity. The loops on the left and right serve as coupling coils. (b) Coupling geometry: The dashed line represents the AC magnetic field of the TE_{011} mode, which is very homogeneous at the location of the Er:YSO crystal (purple). (c) Orientation of the crystal axes.

The cavity is operated in the overcoupled regime with a loaded quality factor of $Q_l = 800$. This corresponds to a cavity linewidth of 3.5 MHz. In oder to determine the internal quality factor, the cavity was measured in the strongly undercoupled regime where Q_l is dominated by the internal quality factor. At low temperature, the internal quality factor is 72300, which is 3.5 times larger than at room temperature. In order to study the AC g-factor, the crystal is rotated around its D_2 axis, which points along the symmetry axis of the cylindrical cavity, see Fig. 5.18(c). For each orientation, the coupling strength of the S_{2b} transition is extracted and the result is shown in Fig. 5.19. The maximum coupling strength measured is $\upsilon_N/2\pi = 21.2 \pm 0.3$ MHz with an inhomogeneous spin linewidth of $\Gamma_2^\star/2\pi = 18 \pm 0.7$ MHz, which proves strong coupling between the Er:YSO spin ensemble and the 3D cavity. The dependence of the collective coupling strength is then compared to a simulation of the AC g-factor using the Matlab package EASYSPIN [SS06] by superimposing both dependencies. The result shows an increase of the coupling strength from 4 MHz at 0° to 21 MHz at 70° originating from a five fold increase of the AC g-factor. A full characterization of the sample can be found in Ref. [Pro+14c].

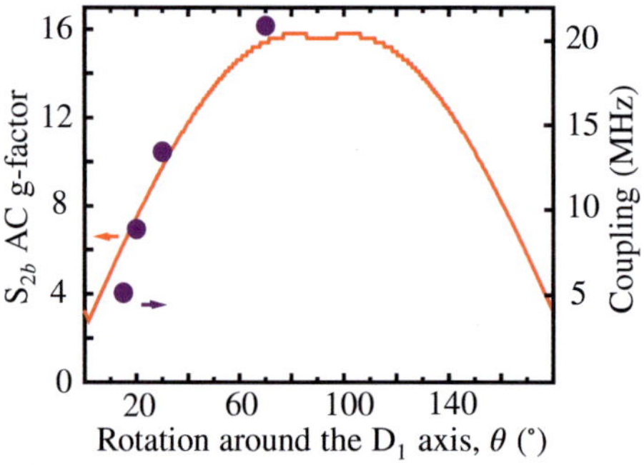

Figure 5.19: Collective coupling strength vs. rotation angle θ. The solid line shows the simulated AC g-factor for the S_{2b} transition. The size of the points covers the approximate uncertainty of the data.

6 Time resolved microwave spectroscopy

In this chapter, the coherence properties of a 50 ppm doped Er^{3+}*:*Y_2SiO_5 *crystal are investigated by pulsed on-chip ESR. In contrast to the previous chapter where the spin ensemble has been studied in the few photon regime, strong* π *and* $\pi/2$*-pulses will be used which manipulate all the spins of the ensemble. The experimental setup is able to rotate the spins by* π *within 40 ns. Here, a copper-based CPW resonator with a loaded quality factor of approximately 133 is employed. This resonator has an effective mode volume of about 11.78 mm*^{3}*, giving rise to a single spin coupling of 2.4 Hz while assuming an AC g-factor of 15. This is more than an order of magnitude smaller compared to the lumped element resonators employed in the previous chapter. This setup can be regarded as a prototype quantum memory, which enables multimode storage of quantum information. We use this test-bed for probing the coherent interaction of* Er^{3+} *with its environment and demonstrate coherent multimode storage of up to 16 excitations with state of the art efficiency.*

6.1 Inhomogeneous spin linewidth and collective coupling revisited

In contrast to the investigations in the previous chapter, a non-superconducting copper CPW resonator is employed. Normal metal resonators have the advantage that they do not perturb the magnetic DC field that is required for the Zeeman shift. Therefore, we expect to measure a minimal inhomogeneous linewidth of the Er^{3+}:Y_2SiO_5 spin ensemble. In the following, the hybrid system is characterized by continuous wave spectroscopy.

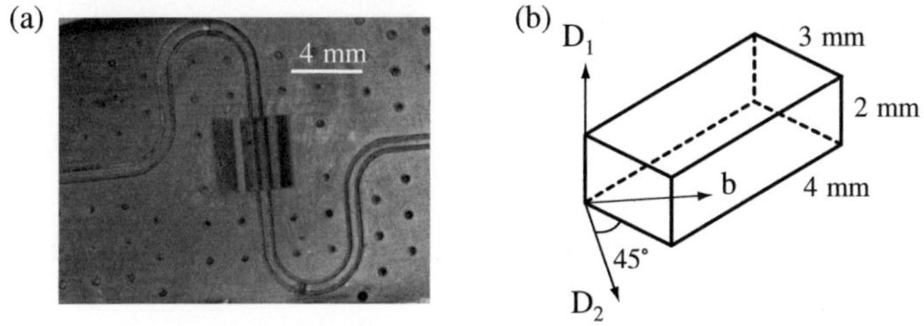

Figure 6.1: (a) Crystal on copper CPW resonator. (b) Orientation of the crystal.

Figure 6.1(a) shows a picture of the sample. The Er:YSO crystal is pressed onto the resonator by a Teflon screw. The orientation of the crystal is displayed in Fig. 6.1(b) and is chosen such that the DC g-factor for the S_{2a} transition becomes minimal and the coupling strength maximal. Since the goal is to apply strong MW pulses later in the experiment, no cryogenic attenuation has been used. Due to the usage of stainless steel cables on the input line, 13 dB attenuation is achieved. On the output side, the last part from 60 K to room temperature is bridged with CuNi cables whose total attenuation amounts for 11 dB. At this point, it is important to discuss the implications of room temperature black body radiation on the sample because the amount of attenuation is very low. In order to estimate the noise power entering the resonator, a frequency band of $\Delta f = 50$ MHz around 3.72 GHz is assumed corresponding to about twice the full resonator bandwidth, which is a conservative estimation. For a matched circuit $P_{noise} = k_B T \Delta f$, which yields for $T = 300$ K a total noise power of ~ -106 dBm at the sample. Thus, the number of photons in the cavity is roughly on the order of ~50. This is not sufficient to saturate the spin ensemble.

The sample is characterized in the usual way by continuous wave MW transmission spectroscopy while sweeping the magnetic field. Figure 6.2 presents the resulting spectrum recorded at 25 mK and a probing power of ~100 fW. The transmitted amplitude is color coded and all four sub-ensembles are resolve with g-factors 14.2, 4.0, 1.9 and 1.1. A clear anticrossing is observed for the high field transition at a magnetic field of 246 mT. Moreover, the electronic transition close to 140 mT also shows a weak anticrossing indicating the onset of strong coupling. The spectrum features a number of weak hyperfine lines, but is was not possible to associate them to specific transitions with sufficient confidence. Specifically, the $^{167}Er^{3+}$ level structure is dominated by the hyperfine interaction because the resonance frequency of 3.7 GHz lies below the maximum zero field splitting of approximately 5.6 GHz. This leads to a number of avoided level crossings and a strong mixing of the states expressed in the uncoupled basis.

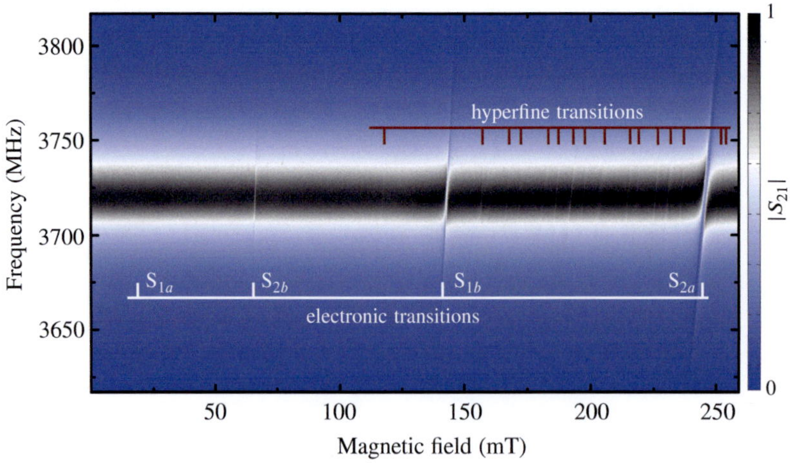

Figure 6.2: $|S_{21}|$ transmission vs. frequency and magnetic field. Four sub-ensembles and a number of hyperfine transitions are visible. The S_{2a} transition shows a clear avoided level crossing.

The following analysis concentrates on the high field transition at 246 mT with a DC g-factor of about 1.1. The avoided level crossing is analyzed by first fitting the resonator far away from the transition. This yields a resonance frequency of ~ 3.721 GHz and HWHM linewidth of $\sim 14.0 \pm 0.1$ MHz. Then, these parameters are kept fixed when fitting the mode splitting. It turns out that this does not fit the lineshape properly. If the resonator linewidth is added as a free parameter, the fit is much better, see Fig. 6.3. We extract a collective coupling strength of $\upsilon_N/2\pi = 13.20 \pm 0.7$ MHz and a HWHM inhomogeneous spin linewidth of $\Gamma_2^\star/2\pi = 7.3 \pm 0.4$ MHz. The fitted resonator linewidth is $\kappa_l/2\pi = 8.2 \pm 0.5$ MHz being a factor of 1.7 smaller than the off-resonant linewidth. This increases the loaded quality factor[1] to $Q_l \approx 227$. Currently, we cannot explain this peculiar behavior. Despite this fact, we confirm that the hybrid system is indeed strongly coupled because we observe vacuum Rabi oscillations, see Sec. 6.3. The change of the linewidth may be attributed to the fact that we couple a microwave resonator characterized by a *homogeneous* linewidth to a spin ensemble forming an effective harmonic oscillator with an *inhomogeneous* linewidth. Moreover, it should be noted that this is a very interesting parameter regime where $\kappa_l > \Gamma_2^\star$, which has not been explored so far in the research on hybrid quantum systems. Typically, superconducting resonators are employed with sub-MHz linewidths exceeding the inhomogeneous linewidths of spin ensembles by orders of magnitude. Therefore, a detailed theoretical study is required to model the influence of the inhomogeneous broadening entirely. According to the fit, κ_l and $\Gamma_2^\star$ are of the same order of magnitude, placing the system close to the so-called impedance matched regime [Afz+13]. This is also supported by the observation that the signal strength of the mode splitting is on the same order of magnitude as the bare resonator transmission, see Fig. 6.3.

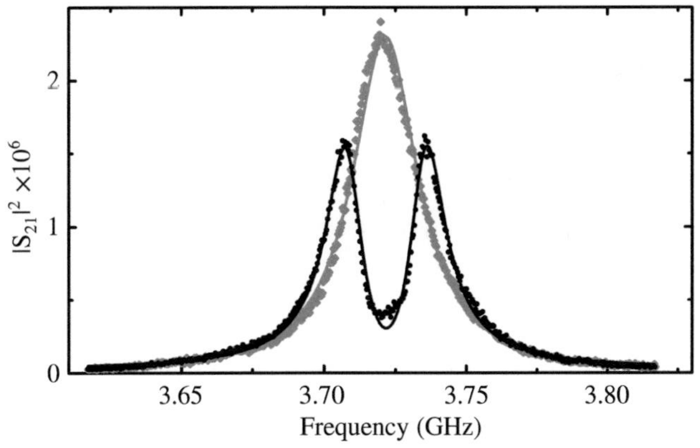

Figure 6.3: Avoided level crossing of the resonance at 246 mT (black points). The solid black line represents a fit to the data. For comparison, the resonator at 227.7 mT is displayed (gray dots).

[1] Since this measurement is performed in transmission, we cannot directly determine the internal quality factor Q_i. However, from experiments where we tried to maximize Q_l, we estimate that Q_i is $\gtrsim 400$

Since this Er:YSO sample has a four times lower ion concentration than the samples investigated in the previous chapter, it is interesting to compare the spin-lattice relaxation times. A 3 s long pulse with a power of 1 μW saturates the spin ensemble, which is subsequently probed at a frequency of 3.72 GHz and a MW power of 10 nW. At high temperatures, the sample operates in the weak coupling regime such that the spin-lattice relaxation time can be determined by fitting an exponential decay curve to the $|S_{21}|$ data. This is justified until 0.45 K because below, the decay curves deviate from pure exponentials. Figure 6.4(a) shows the obtained T_1 values from 0.45 K to 1.0 K. According to theory, the dominant energy relaxation mechanism is a direct process (see Sec. 2.3.5) which scales as $T_1^{-1} \propto \coth(\hbar\omega/2k_BT)$. Since the resonance frequency $\omega/2\pi$ is known, the only free parameter is the proportionality constant. The purple line in Fig. 6.4(a) displays the fit of this function to the data. By extrapolation to 30 mK, a T_1 of approximately 100 s is predicted.

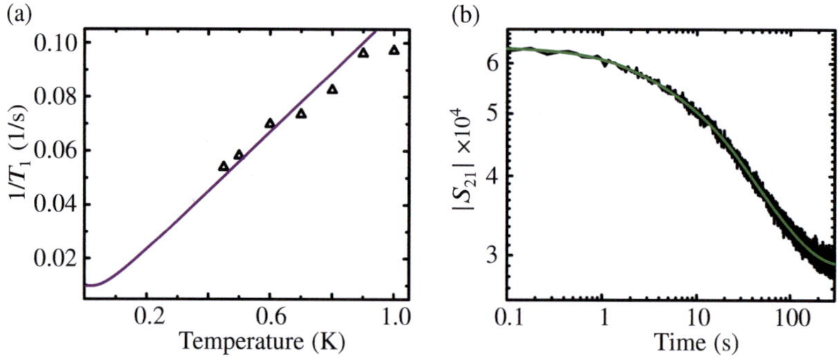

Figure 6.4: (a) Temperature dependence of the spin-lattice relaxation time. The purple line is a fit assuming a direct relaxation process. (b) $|S_{21}|$ decay curve after a saturation pulse, plotted in double-logarithmic scale. The decay strongly deviates from a simple exponential decay curve. The green line is a fit employing a sum of three exponential decays. After 100 s, $\gtrsim 90\%$ of the spins are relaxed back to equilibrium.

In contrast, Fig. 6.4(b) presents a $|S_{21}|$ decay curve measured at 30 mK and plotted with double-logarithmic scaling. The decay strongly deviates from a simple exponential decay. Also, it is not possible to fit the signal employing the strong coupling theory, see Eq. (3.35) in Sec. 3.3. This may originate from the fact that the bare loaded quality factor κ_l depends on whether the cavity is coupled to the spin ensemble, see the analysis presented in Fig. 6.3. The only function which achieves a perfect fit to the data is a sum of three exponential decays, represented by the green line in Fig. 6.4(b). The decay times are 2.76 ± 0.26 s, 19.2 ± 1.1 s and 73.9 ± 1.7 s. After 100 s, the signal has decayed by $\gtrsim 90\%$, indicating that most spins are in thermal equilibrium. The dominant decay is given by the second timescale of the fit function, yielding an approximate $T_1 \approx 20$ s. At this point we have to conclude that the details of the relaxation process appear to be much more complex, such that we can only speculate that the strongly coupled cavity offers an additional energy decay channel for the spin system.

The hybrid system operates in an interesting regime, where $\kappa_l > \Gamma_2^\star$. To investigate the dependence of the power spectrum $|S_{21}|^2(f)$ on the inhomogeneous spin linewidth $\Gamma_2^\star$, an additional magnetic field gradient is applied.

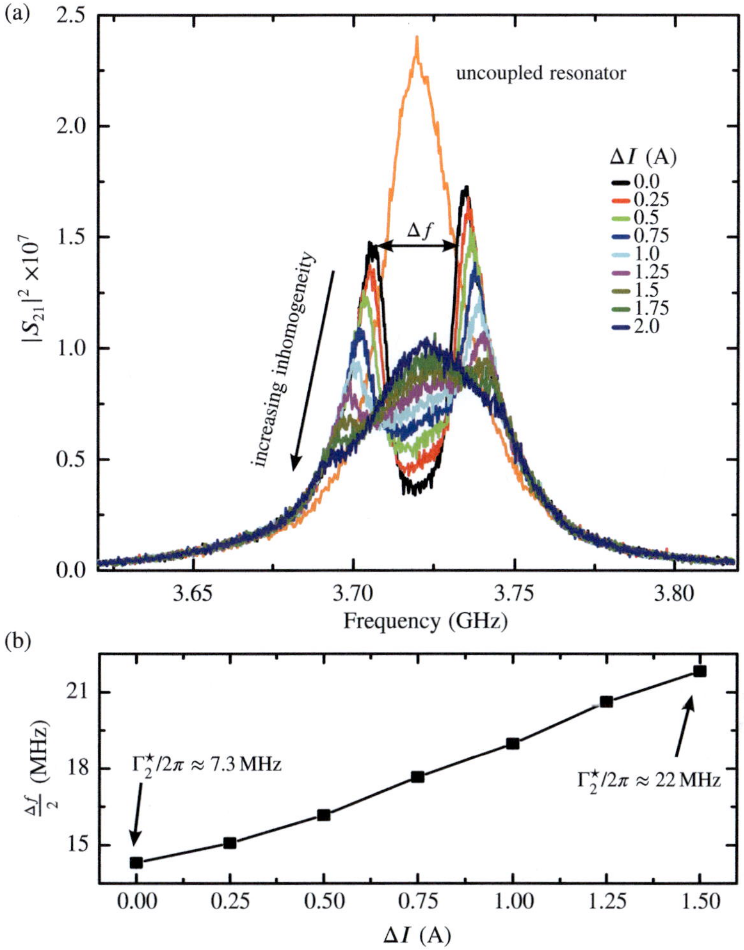

Figure 6.5: (a) Power spectra of the Cu resonator coupled to the spin ensemble. The colors indicate spectra with different magnetic field inhomogeneity introduced by the Helmholtz coil pair. For comparison, the orange line presents the spectrum of the uncoupled resonator. (b) Half mode splitting vs. electrical current inhomogeneity $\Delta I = (I_1 - I_2)$ in the Helmholtz coil pair.

This is achieved by driving the two coils of the Helmholtz coil pair with different currents I_1 and I_2. Figure 6.5(a) presents the resulting spectra of the resonant hybrid system. The colors denote power spectra at different inhomogeneity $\Delta I = (I_1 - I_2)$. The average current is kept fixed to $(I_1 + I_2)/2 = 7.121$ A. For comparison, the orange line shows the $|S_{21}|^2(f)$ transmission of the uncoupled resonator measured at a different magnetic field away from the spin transition. In the absence of magnetic field inhomogeneity (black line), a clear mode splitting is observed. As ΔI increases, the amplitude of the peaks of the mode splitting decreases. Simultaneously, the separation Δf of the peaks of the mode splitting increases. In the center of the mode splitting, the amplitude increases, such that above $\Delta I = 1.25$ a three peak structure emerges.

Such a complex behavior is not covered by the theory of two coupled quantum harmonic oscillators given by Eq. (3.25). Therefore, most of the spectra cannot be fitted with confidence. However, one can draw important conclusions from the peak separation of the mode splitting. Namely, in first order approximation, the half width of the peak separation yields the collective coupling strength $\frac{\Delta f}{2} \approx \upsilon_N$. Figure 6.5(b) presents the result. With increasing inhomogeneity, the peak separation grows, which suggests that the collective coupling strength might increase. This is very peculiar because the collective coupling strength is typically assumed to be independent of the inhomogeneous broadening!

On the scale of the crystal's dimensions, the gradient magnetic field is approximately linear and causing a small deviation of the precession frequency proportional to the position of the spin in the crystal. Thus, the resulting inhomogeneous lineshape is a convolution of the intrinsic inhomogeneous lineshape of the ensemble and a box-like shape resulting from the gradient field. Above $\Delta I = 1.5$ A, the peaks of the mode splitting vanish. So, the spectrum at $\Delta I = 1.5$ A is situated at the border of the strong and weak coupling regime. The modified inhomogeneous spin linewidth at $\Delta I = 1.5$ A has been calculated by using a magnetic field simulation of the Helmholtz coil pair yielding $\Gamma_2^\star \approx 22$ MHz. This value is on the same level as $\Delta f/2 = \upsilon_N$ and larger than the bare resonator linewidth. A complete understanding of the underlying effects requires a complex theoretical treatment similar to Ref. [Put+14; Kri+14] and to our knowledge, the regime where $\upsilon_N \approx \kappa_l \approx \Gamma_2^\star$ has never been explored so far.

6.2 Time domain setup

In order to measure the coherence properties of the spin ensemble, time resolved measurements are necessary. This imposes a considerable challenge on the experimental setup. The single spin coupling is approximately seven orders of magnitude smaller than the collective coupling strength. To perform a π rotation of the whole ensemble within the coherence time, a high power microwave pulse is needed because the Rabi frequency is proportional to the single spin magnetic moment $\Omega = g\mu_B B_{AC}/2\hbar$.

On the other hand, the setup should guarantee a good isolation such that no microwave power leaks into the system between the pulses. This is crucial when the hybrid system is operated in the strong coupling regime where the collective $\sqrt{N}$ enhancement of the coupling strength makes the system extremely susceptible to low microwave powers resulting in a partial saturation of the spin ensemble. The collective coupling strength has been determined in the previous chapter $\upsilon_N/2\pi = 13.2$ MHz. The vacuum Rabi frequency is given by $\Omega_0 = 2\upsilon_N$

such that a single photon is transferred to the spin system within $\sim$ 18.9 ns due to the collective coupling. In contrast, to rotate all spins within 40 ns in the excited state (π-pulse), 1.6×10^{14} photons are required.

Figure 6.6 presents a schematics of the setup. The microwave pulse generation with a large on/off ratio is achieved by a two step process. A two-channel Tektronix AWG 7062 arbitrary waveform generator (AWG) generates two pulses of equal length (10-40 ns) but of different shape. The first one is a rectangular pulse switching the output of an Agilent E8257C MW source on for the duration of the pulse. The microwave frequency of the source is detuned by 200 MHz below the microwave resonator frequency. The generated microwave pulse enters the local oscillator port of a frequency mixer. There, the frequency of the pulse is shifted by 200 MHz to match the frequency of the microwave resonator. The required 200 MHz IF signal is generated by the second channel of the AWG. Then, the signal is amplified by 30 dB and sent into the cryostat.

Note that only for the duration of the pulse, the microwave is tuned into resonance with the microwave resonator. This means that parasitic microwave power leaking into the cryostat in between the pulses is off-resonant and does not perturb the sample. To achieve the required phase stability, all devices are connected by a common 10 MHz reference clock. Note that the mixer shown in Fig. 6.6 generates two sidebands but only the upper sideband is resonant with the sample whereas the lower sideband is 400 MHz detuned and cannot perturb the sample. The attenuator shown in the schematics represents all attenuation along the input line including losses at the mixer. In a typical experiment, the output power of the Agilent microwave source is set to 0 dBm and further amplified by 30 dB. Taking all attenuation in the system into account, the pulse has an approximate power of 10 mW at the sample.

The transmitted signal is first amplified and filtered by a high-pass filter before it gets demodulated by an IQ mixer. The local oscillator of the IQ mixer is supplied by an Anritsu M637022A source, which has the same frequency as the Agilent source used for the pulse generation. Both sources are frequency locked. Since the local oscillator is detuned from the frequency of the incoming microwave pulse, a 200 MHz signal remains on the I and Q port. This signal is low-pass filtered and then acquired by a Tektronix TDS5054B oscilloscope. The input ports of the oscilloscope operate in the AC mode to get rid of DC offsets. In a final step, the data is transferred to a standard personal computer where the 200 MHz is removed from the I and Q signal by a digital mixer. This mixer is implemented by multiplying the signal with a local oscillator at frequency $f_{\mathrm{LO}} = 200$ MHz and subsequent filtering using a second order digital Butterworth filter [But30] with a cutoff frequency of $f_{3\mathrm{dB}} = \frac{1}{4} f_{\mathrm{LO}}$. By recording a 200 MHz signal instead of a DC pulse, DC offsets and low frequency noise including drifts are avoided. This gives a much better signal-to-noise ratio and makes the data acquisition reliable and robust.

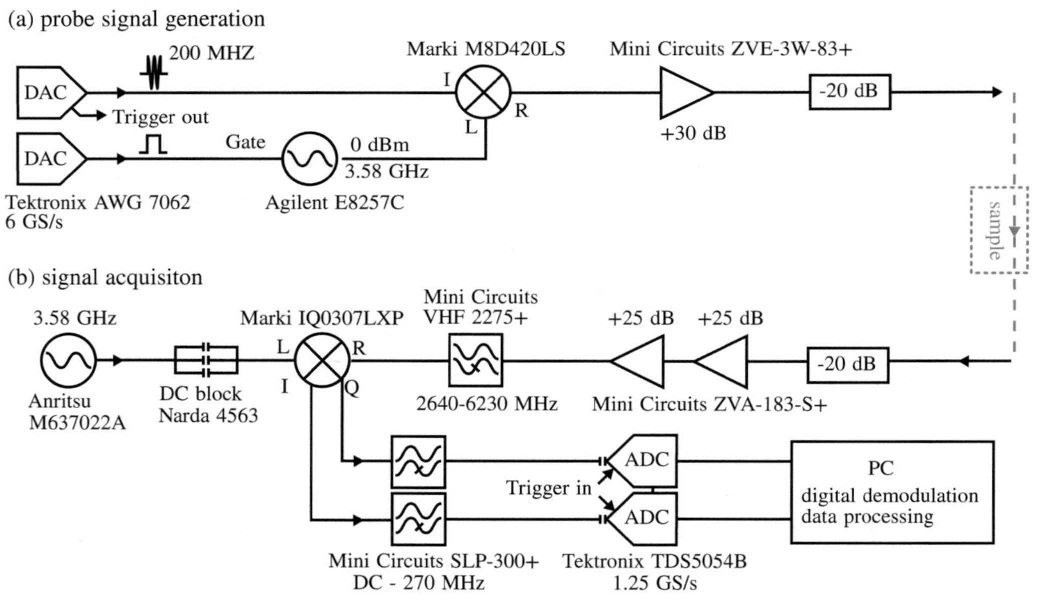

Figure 6.6: Time resolved ESR spectroscopy. (a) Probe signal generation: The MW excitation pulses are generated in two steps by fast switching of a MW source and subsequent 200 MHz modulation using a mixer. (b) Signal detection: The MW signal is amplified, filtered and mixed down to 200 MHz. The signal is digitized and digitally mixed down to DC by a PC. All devices are synchronized by a common 10 MHz reference clock (not shown).

6.3 Vacuum Rabi oscillations

In the strong coupling regime, the resonant MW cavity and spin oscillator energy states are no longer eigenstates of the coupled system. Instead, they hybridize and the coupling lifts the degeneracy of the resonant energy levels by the vacuum Rabi splitting. The vacuum Rabi frequency is given be $\Omega = 2\upsilon_N$. This means that a photon in the resonator will get transferred to the spin system and back to the resonator in a time $\tau_R = 2\pi/\Omega$. For our system with a collective coupling strength of $\upsilon_N/2\pi = 13.2\,\text{MHz}$, this Rabi oscillation period amounts to $\tau_R = 37.9\,\text{ns}$.

It is possible to observe the coherent dynamics of the coupled cavity-spin systems already in the low MW power regime with pulse lengths that exceed the coherence times of the coupled system. This was first experimentally demonstrated for a hybrid circuit QED system consisting of an NV spin ensemble coupled to a SC CPW resonator [Put+14]. In our experiment, we applied a long ($\Delta t_p \gg 1/\kappa, 1/\Gamma_2^\star$) and resonant MW pulse exciting the hybrid system and driving it into the steady state. Simultaneously, we observe the transmitted power as a function of time. When the system is in the steady state, the MW drive is switched off. Figure 6.7(a) shows the recorded microwave power vs. time of the whole pulse sequence. Both at the beginning and the end of the pulse, an oscillation is observed. These occur due to the coherent exchange of photons between the resonator and the spin system. The radiated power is directly proportional to the energy stored in the cavity. An extended theoretical analysis of the dynamics can be found in Ref. [Kri+14]. Since the collective coupling strength is not much larger than the decay rates of the hybrid system, only one clear oscillation period is observed. Figure 6.7(b) presents a zoom at the end of the pulse. Superimposed on the exponential decay of the microwave signal from the cavity, a clear oscillation is observed. To determine the oscillation period, the time difference between the first minimum and maximum is measured and multiplied by two yielding $\tau_R = 66.4\,\text{ns}$.

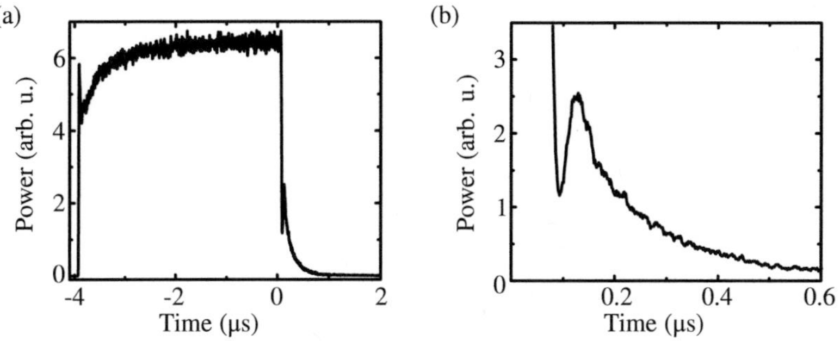

Figure 6.7: (a) Transmitted resonant microwave power vs. time. The beginning and end of the pulse shows a characteristic oscillation. (b) Zoom onto the oscillation at the end of the pulse.

This disagrees with the value obtained from the cw spectroscopy. It would correspond to a coupling of only $\tilde{\upsilon}_N/2\pi \approx 7.5\,\text{MHz}$. This places the system at the transition region between strong and weak coupling considering $\kappa_l/2\pi = 8.2\,\text{MHz}$ and $\Gamma_2^\star/2\pi = 7.3\,\text{MHz}$. The small coupling can be explained by noting that the experiment has been performed with a rather large microwave power of approximately 1 mW at the sample applied for 4 μs. This saturates the spin ensemble considerably such that the effective coupling strength given by $\tilde{\upsilon}_N \propto \sqrt{N_\downarrow - N_\uparrow}$ is reduced. Here, the arrows denote the spins in the ground and excited state, respectively. From this, we can infer the amount of polarization assuming full polarization in the cw measurement

$$P = \frac{N_\downarrow - N_\uparrow}{N_\downarrow} = \frac{\tilde{\upsilon}_N^2}{\upsilon_N^2} \approx 32.3\% \,. \tag{6.1}$$

However, the straightforward way of measuring vacuum Rabi oscillations is to send a short MW pulse into the cavity and measure the emitted microwave power response. In this experiment, the MW frequency was fixed to the cavity frequency at 3.719 GHz and the magnetic field was swept in order to investigate the two high field transitions and compare the results to the off-resonant case. Figures 6.8(a) and (b) display the response of the hybrid system to a 20 ns MW pulse with a power of approximately 10 mW corresponding to about 81×10^{12} photons. In each plot, the gray line denotes the off-resonant response and the black data represents the resonant response.

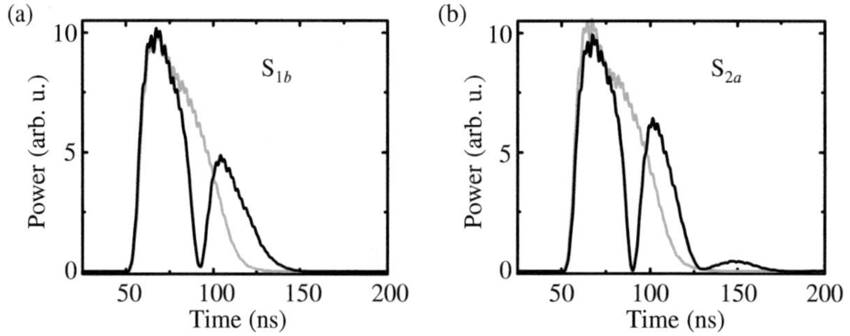

Figure 6.8: Response to a 20 ns MW pulse. The grey line shows an off resonant pulse for comparison. (a) S_{1b} transition (143.0 mT). (b) S_{2a} transition (245.8 mT).

The weaker coupled transition S_{1b} exhibits longer oscillation periods, see Fig. 6.8(a). The signal of the high field transition S_{2a} presented in Fig. 6.8(b) shows one more oscillation. The oscillation period measured from the first minimum to the second minimum amounts for $\tau_R = 39.8$ ns being in good agreement with the value obtained from the cw spectroscopy $\tau_R = 37.9$ ns. As we will see in the next section, the pulse employed in this experiment corresponds to an effective $\pi/2$-pulse. Thus, a large amount of energy is transfered into the system but apparently the Tavis-Cummings physics still seems to apply. It would be interesting to repeat the same experiment in the single photon regime which is currently not possible due to the insufficient sensitivity of the setup. A suitable setup should consist of a cryogenic Josephson parametric amplifier [Ber+10] with a subsequent cryogenic HEMT amplifier to achieve a reasonable signal-to-noise ratio.

Interestingly, the resonant pulse (black line) shows for both transitions a longer decay time compared to the off-resonant case (gray line). Since the decay rate of the spin system caused by the transfer of excitations into dark modes is smaller, the photons remain longer in the hybrid system. In other words, the spin ensemble extends the lifetime of the photon and therefore serves as certain kind of memory here.

6.4 Spin coherence properties

Figure 6.9 shows a typical two pulse echo (2PE) sequence. An initial 20 ns long and 10 mW strong $\pi/2$-pulse rotates the spin ensemble to the equatorial plane where it starts dephasing. A subsequent 40 ns long 10 mW π-pulse applied after a delay time $\tau = 400$ ns reverses the time evolution and the ensemble emits an echo at time 2τ.

The recorded pulse shapes of the $\pi/2$ and π-pulse feature vacuum Rabi oscillations[2]. Although the $\pi/2$ pulse width is only 20 ns, the width of the echo is on the order of 100 ns due to the finite bandwidth of the microwave cavity. The amplitude of the echo is proportional to the $\langle \hat{S}_y \rangle$ expectation value of the spin ensemble, see Sec. 2.3.6. In general, the 2PE sequence refocuses dephasing introduced by the inhomogeneous frequency broadening, such that it is possible to determine the spin coherence time T_2 by measuring the echo amplitude as a function of twice the delay time 2τ.

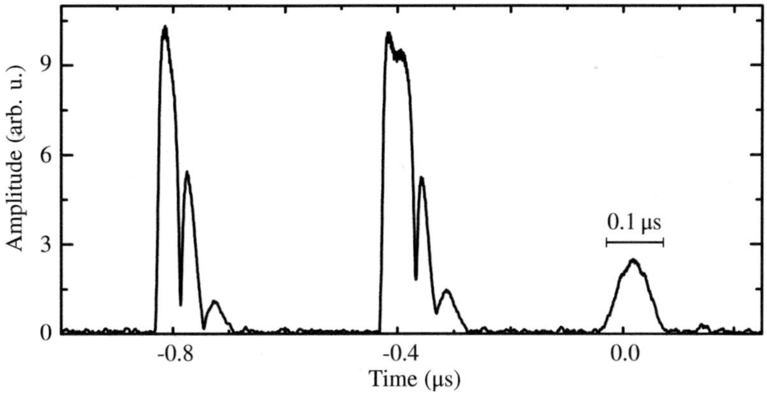

Figure 6.9: Spin echo sequence with a pulse separation of $\tau = 400$ ns. Each pulse is accompanied by vacuum Rabi oscillations. The $\pi/2$-pulse has a width of 20 s, but the echo appears as a 100 ns wide pulse of approximately Gaussian shape.

The echos are analyzed by integrating over the full echo power signal to obtain the total radiated energy. Then, the square root of the energy is plotted vs. 2τ and fit to an exponential decay to retrieve the T_2 time. Figure 6.10 presents such a T_2 decay curve recorded at a temperature

[2] Note that the free induction decay is not observable here because the large inhomogeneous broadening results in a very short T_2^* time of 22 ns.

of 30 mK. The signal features prominent oscillations and the solid line represents a fit to an exponential decay including an oscillatory term. This yields a T_2 time of approximately 5.6 μs.

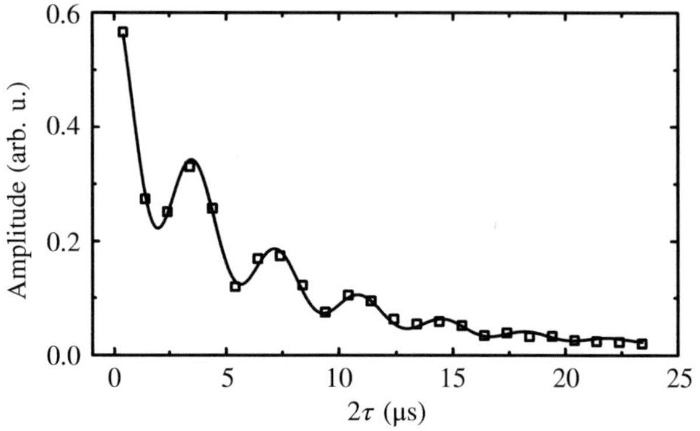

Figure 6.10: The exponential echo decay is modulated by prominent oscillations. The solid line presents a fit to the data with a T_2 time of 5.6 μs.

The oscillatory modulation of echo decay curves is called *electron spin echo envelope modulation* (ESEEM), Sec. 2.3.6. It originates from the dipole-dipole coupling of the Er^{3+} spin to its coherent magnetic environment. The closest magnetic dipoles are the nuclear spins of the yttrium ^{89}Y ions surrounding the erbium ion at a minimal distance on the order of 4 Å. The dipole-dipole coupling gives rise to a super-hyperfine splitting of the electronic spin states by two super-hyperfine frequencies. A theoretical treatment of this system is provided in Sec. 2.3.6. To investigate this effect with higher fidelity, a different pulse sequence is employed, which allows for observing the modulations over a longer timescale.

The stimulated pulse echo sequence (3PE) consists of three $\pi/2$ pulses and measures the phase memory T_m. In the absence of spectral diffusion, it probes the longitudinal relaxation time. In spin systems, the phase memory time T_m is typically longer than T_2 but smaller than the energy relaxation time T_1. Therefore, the 3PE sequence is well suited to study ESEEM effects with higher resolution. Figure 6.11 shows the 3PE echo decay. The plot shows clear oscillations at the beginning, which fade out and reappear at $T_W \approx 40$ μs. This beating is a clear sign for the presence of the two super-hyperfine frequencies.

The decay curve can be fit best with a double exponential decay with characteristic timescales of 2.1 ± 0.4 μs and 23.9 ± 1.2 μs. The fitted double exponential decay is then used to normalize the data. After normalization, the data is Fourier transformed using a Hamming window to suppress numerical artifacts. The Fourier power spectrum presented in Fig. 6.12(a) shows a clear broad peak at about 500 kHz. The nuclear spin of ^{89}Y has a g-factor of -0.2748 [Gui+07], which yields a free precession frequency in the experimental field of $\sim$ 246 mT of 515 kHz. This is in good agreement with the experiment confirming that it is indeed the ^{89}Y nucleus interacting with the erbium spin. Due to the limited time resolution of the 3PE experiment, the Fourier spectrum does not resolve the two hyperfine frequencies.

Additionally, the measured Fourier spectrum is compared to a numerical simulation, where the four closest Yttrium ions were considered to contribute most, see Fig. 6.12(b). The crystallographic data from Ref. [MBB67] has been used to measure the ion-ion distances and angles with respect to the magnetic DC field of the experiment. Due to the small DC g-factor, the dipole-dipole interaction is small. Therefore, the super-hyperfine frequencies are too close to be resolved in the experiment. Note that it is the electronic spin coherence which limits the frequency resolution.

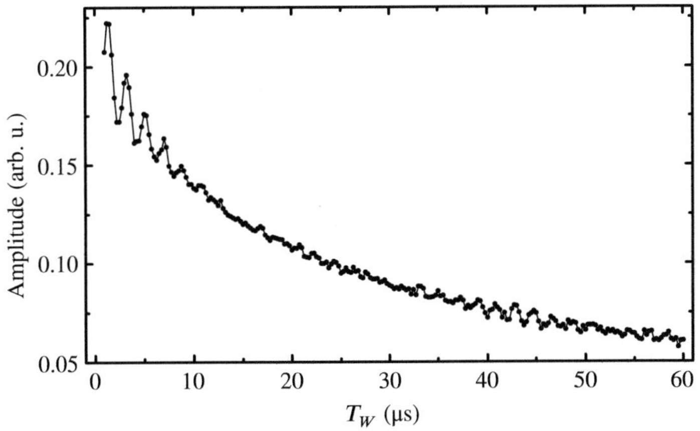

Figure 6.11: Stimulated echo decay. The beating between the two hyperfine frequencies is clearly visible. The data can be fit with a double exponential decay with decay times of 2 and 23.9 µs.

The presence of a super-hyperfine interaction may have an interesting consequence. The Er^{3+} ion could act as a local spin bus to address individual ^{89}Y nuclear spins in its vicinity, which may perform subsequent computations. However, the current electronic spin coherence times limit the frequency resolution such that in the present experiment, the ESEEM frequency resolution is not sufficient to distinguish individual nuclear spins. Therefore, operation in a regime with larger DC g-factor is required. This enhances the super-hyperfine interaction and separates the individual nuclear spins from each other in the Fourier spectrum. A minimum frequency separation exceeding the 35 Hz homogeneous linewidth of the ^{89}Y nuclear spins is required [DS88].

In this context, it is interesting to look at the impact of the super-hyperfine interaction on the homogeneous broadening of the Er spin, which gives rise to the so-called *frozen core effect* [Hul05]. An important source of decoherence for the Er spin are nuclear spin flips because, at milli-Kelvin temperatures, the nuclear spin bath is not yet polarized by the magnetic fields employed in this experiment. A nuclear spin-flip must conserve energy such that the dominant process is the simultaneous flipping of two or more nuclear spins at a time (flip-flop process). The Er^{3+} ion produces a magnetic field in its surrounding, which shifts the energy of the ^{89}Y nuclear spins in its vicinity. Due to this energy discrepancy with the rest of the nuclear spin bath, it is harder to meet the energy conservation requirement for those ^{89}Y nuclear spins around the Er ion. Thus, nuclear spin-flip processes are suppressed in the vicinity of the erbium ion improving its coherence. Note that the stability of the frozen core improves with increasing DC g-factor inducing a larger energy discrepancy. In the presented experiment, the DC g-factor is about 1.1, which gives rise to only a small shift of the nuclear precession frequencies. Since spin-flip processes are not entirely suppressed, the frozen core will *melt* on longer time scales. With the appearance of the first nuclear spin-flips, the homogeneous broadening increases as a function of time giving rise to spectral diffusion, see Sec. 2.3.4.

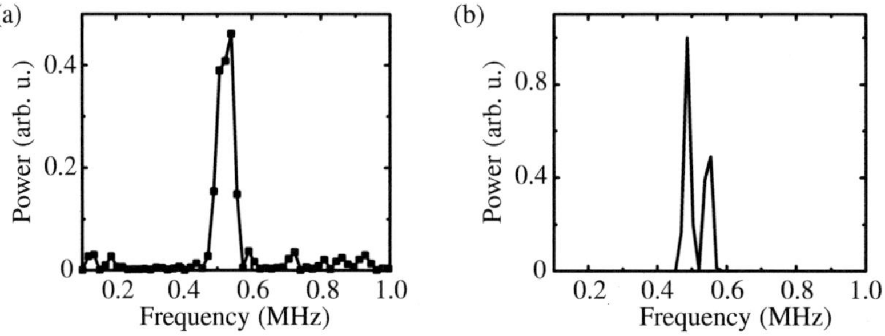

Figure 6.12: (a) FFT of the ESEEM power spectrum. (b) Simulation of the frequency spectrum based on the crystal structure.

Decoherence induced by the electronic spin bath

There are two major classes of decoherence: (I) Dephasing induced by the magnetic environment of the spin ensemble, and (II) direct spin-spin interaction within the spin ensemble. Source (I) can be addressed by investigating the temperature dependence of the coherence time T_2. Figure 6.13(a) presents the temperature dependence of the coherence time from 30 mK up to 1 K. Note that the vertical axis of the plot is given in $1/T_2$ and both axes are scaled logarithmically for improved visibility. In the displayed temperature range, the T_2 time decreases by approximately 1 μs. Below 50 mK, T_2 remains constant. This indicates that magnetic fluctuations are frozen out at these temperatures. In order to interpret the data, it is important to remember that Er:YSO consists of four sub-ensembles. Those present a considerable magnetic background. From the continuous wave microwave spectroscopy, the g-factors for all sub-ensembles were determined. Then, we calculate the Zeeman shift of all sub-ensembles at the magnetic field of the experiment. Figure 6.13(b) shows the corresponding transition frequencies. The lowest frequency of 3.72 GHz belongs to the S_{2a} ensemble, which is the sub-ensemble under investigation. The remaining three ensembles in the frequency range from 6.39 GHz to 48.81 GHz represent a fluctuating magnetic background. To put the transition frequencies into perspective, Fig. 6.13(b) also provides the effective Zeeman temperatures $T^{(z)} = \hbar\omega/k_B$, where k_B is the Boltzmann constant. The lowest Zeeman temperature of 307 mK belongs to the S_{1b} ensemble. As the temperature rises, the background spins perform random spin-flips altering the precession frequencies of the S_{2a} ensemble under investigation.

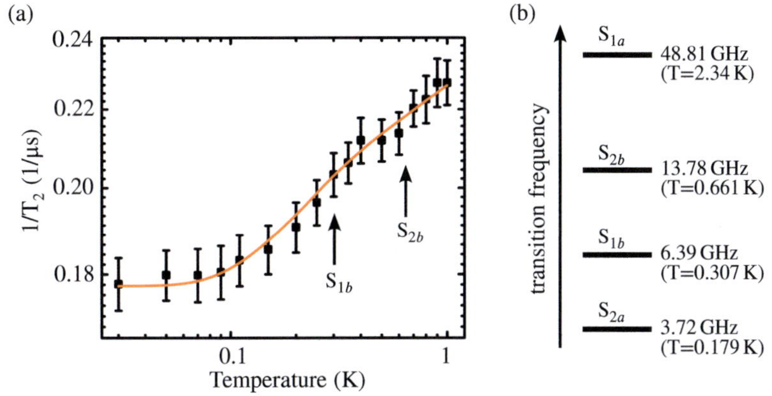

Figure 6.13: (a) $1/T_2$ vs. temperature. Note the logarithmic scale. Between 30 mK and 1 K, the T_2 time decreases from approximately 5.6 μs to 4.4 μs. The solid line is a fit to theory considering decoherence induced by thermal fluctuations from the surrounding sub-ensembles. (b) Transition frequencies and effective temperatures at 246 mT of all sub-ensembles.

Specifically, spin flip-flop processes are the dominant flipping mechanism. These processes become important in relatively concentrated samples [KA62] and consist of a spin flipping up while another one is flipping down or vice versa. Consequently, this rate is proportional to the product of the occupation probabilities of the spin up and the spin down state for a given temperature. The occupation probabilities are given by the Boltzmann statics (Sec. 3.2.5), $P_\uparrow = [1 + \exp(T^{(z)}/T)]^{-1}$ and $P_\downarrow = [1 + \exp(-T^{(z)}/T)]^{-1}$. By including a residual dephasing rate Γ_{res}, the temperature dependence of T_2 is described by [Kut+95; Tak+08]

$$\frac{1}{T_2(T)} = \Gamma_{\text{res}} + \sum_{i\in\{S_{1b},S_{2b},S_{1a}\}} \frac{\xi}{\left(1+e^{T_i^{(z)}/T}\right)\left(1+e^{-T_i^{(z)}/T}\right)}. \tag{6.2}$$

Here, ξ is a temperature-independent free parameter. The orange line in Fig. 6.13(a) presents the fit of Eq. 6.2 to the data. Note that Γ_{res} and ξ are the only free parameters because the Zeeman temperatures are known from the cw spectroscopy. Below 100 mK, the temperature dependence saturates and the fit yields a $\Gamma_{\text{res}} \approx 1/5.63\ \mu\text{s}^{-1}$. Also, the fit nicely reproduces the data including the shallower slope towards larger temperatures. Therefore, we conclude that spin flip-flop processes of the other three sub-ensembles dominate the temperature dependence of the decoherence rate.

The second source of decoherence are spin flip-flop processes within the sub-ensemble under investigation. Those give rise to instantaneous spin diffusion (see Sec. 2.3.4), which cannot be refocused by a two pulse echo sequence (2PE). However, it is possible to measure this dipolar interaction with a modified 2PE sequence where the angle of the second pulse θ_2 is varied from 0 to π. The latter angle would correspond to a normal 2PE sequence. The $\theta_2 = \pi$-pulse typically refocuses the magnetic field inhomogeneity and low frequency noise, which is constant on the timescale of the pulse sequence. However, if a spin flip-flop occurs, the

θ_2-pulse cannot refocus this interaction because the pulse flips both spins involved in the interaction as they belong to the same spin ensemble. So, the flip-flops reduce the observed T_2 time.

On the other hand, if the θ_2-pulse is shortened, it will only refocus a subset of the spins. This can be regarded such that the shortened θ_2-pulse mimics a homogeneously dilute spin ensemble [Bro+11; KA62; SE01]. Figure 6.14(b) shows a sketch of the pulse sequence. In our experiment, the θ_2-pulse is shortened by decreasing the pulse height, and Fig. 6.14(a) presents the resulting reciprocal T_2 times as a function of the θ_2-pulse height. As the pulse height is reduced, $T_2 = 5.6$ µs stays constant until it increases below pulse heights of 50%. In the limit of zero pulse height, T_2 reaches approximately 7 µs. In contrast to literature, the coherence time does not increase immediately when the second pulse is faded out. This can be explained by considering the coupling geometry of the experiment, which has been neglected so far. Figure 6.14(c) presents a schematics of the setup. The closed blue lines represent the magnetic AC probe field of the pulse. To first order, the central conductor of the CPW resonator can be approximated by a wire such that the magnetic field decays proportional to the inverse of the distance r. Since the Rabi frequencies of the spins are proportional to the magnetic field strengths, they scale proportional to $1/r$ as well. Therefore, only in a narrow region indicated by the green area, the condition for a π-rotation is fulfilled. In the rest of the crystal, the mismatched phases contribute weakly to the overall signal, such that the decay curve is dominated by the signal from spins in the green region. By decreasing the pulse height, the magnetic field strength decreases and the green area moves towards the CWP resonator. As the effective area exits the crystal, none of the spins can be refocused efficiently and the measured T_2 time increases.

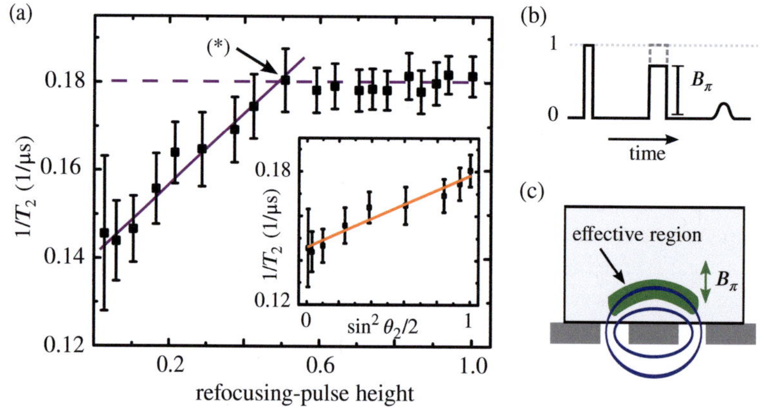

Figure 6.14: (a) By decreasing the height B_π of the refocusing pulse, T_2 increases up to 7 µs. Note that the vertical axis has a reciprocal scale. Interestingly, T_2 stays constant until the field amplitude of the refocusing pulse is reduced by $\sim$ 50%, indicated by (*). The purple lines are guides to the eyes. (Inset) Linear fit to the slope assuming (*) marks the position of the π-pulse, see text for details). (b) Schematics of the pulse sequence. (c) Possible explanation of the observed behavior. Due to the anisotropic AC magnetic field, the region inside the crystal where the Rabi frequency allows a perfect π-pulse depends on the pulse strength.

In order to analyze the data, the point where the slope starts, indicated by the arrow in Fig. 6.14(a), is defined as effective *global* ($\theta_2 = \pi$)-pulse. The data, rescaled to $\sin^2(\theta_2/2)$ of the rotation angle θ_2 of the normalized pulse, is shown in the inset. According to the theory of instantaneous spin diffusion (see Sec. 2.3.4), $1/T_2 \propto \sin^2(\theta_2/2)$. The orange line in the inset of Fig. 6.14(a) shows a linear fit to the rescaled data. There, T_2 varies from 7 µs to 5.6 µs. From the slope, we extract a spin concentration of $C = 1.28 \times 10^{17}\,\mathrm{cm}^{-3}$. For comparison, the concentration given by our manufacturer is 50 ppm corresponding to $\sim 9.35 \times 10^{17}\,\mathrm{Er}^{3+}/\mathrm{cm}^3$. Since we are considering a single sub-ensemble, only 1/4 of the spins participate yielding $C = 2.34 \times 10^{17}\,\mathrm{cm}^{-3}$. This deviates merely by a factor of about 2 from the experimental result confirming that despite the inhomogeneous pulses, the measurement yields a reasonable result.

Using the measured spin concentration, we can calculate the average dipole-dipole interaction strength. The coupling frequency between two dipoles is given by [Bro+11] (Supplementary material)

$$\upsilon_D = \frac{\mu_0 \mu_B^2 g_a g_b}{4\pi\hbar r_{ab}^3}\left(1 - 3\cos^2(\alpha)\right), \tag{6.3}$$

where g_a and g_b are the g-factors of the two spins and α denotes the angle between the direction of the Zeeman field and the inter-spin axis. The parameter r_{ab} quantifies the mean distance of the spins [BC08]

$$r_{ab}(C) = \frac{\left[\Gamma(\frac{3}{2}+1)\right]^{1/3}}{\sqrt{\pi}} \frac{\Gamma(1+\frac{1}{3})}{\Gamma(1)} \left(\frac{1}{C}\right)^{1/3}. \tag{6.4}$$

Here, Γ denotes the gamma function. To determine υ_D for the present ensemble, one has to average over the angular dependence of Eq. (6.3). By inserting the measured spin concentration, a dipole coupling strength of $\upsilon_D/2\pi \approx 11.9\,\mathrm{kHz}$ is determined giving rise to a theoretical limit for T_2 of 13.4 µs. Repeating the calculation with the concentration provided by our supplier yields $\upsilon_D/2\pi \approx 21.6\,\mathrm{kHz}$, which would correspond to a $T_2 \approx 7.4$ µs.

The measured $T_2 = 5.6$ µs determined by the optimal 2PE sequence ($\theta_2 = \pi$) lies below the limits imposed by instantaneous spin diffusion. In fact, the investigation clearly shows that direct dipole coupling within one sub-ensemble can only be made responsible for approximately half of the measured decoherence rate. The remaining contributions may originate from thermal fluctuations of the $^{167}\mathrm{Er}^{3+}$ spin bath. Specifically, the $^{167}\mathrm{Er}^{3+}$ nuclear spin is not fully polarized at the temperature and magnetic field used in the experiment.

In Sec. 5.1, it has been noted that it is not possible to enter the strong coupling regime at high DC g-factors with the 200 ppm Er:YSO2 sample. In this measurement, the low field transition exhibits a g-factor of 13.6 and an inhomogeneous spin linewidth of $\Gamma_2^\star \approx 23\,\mathrm{MHz}$. With our present result, it is now possible to estimate the dipole-dipole interaction. The dipole-dipole interaction υ_D scales linearly with the concentration but quadratically with the DC g-factor.

For a concentration of 200 ppm and a g-factor of 13.6, the dipole-dipole interaction amounts for $v_D \approx 13$ MHz. Thus, dipole-dipole coupling is responsible for half of the line broadening. This is a factor of 611 larger than in the 50 ppm and $g = 1.1$ sample investigated in this chapter.

As a concluding remark, the problem of the anisotropic AC field encountered in these measurements due to the planar architecture has been addressed recently in the literature. So-called adiabatic pulses like WURST-20 (Wideband, Uniform Rate, Smooth Truncation 20) and BIR-4 (B_1 Insensitive Rotation 4) [Sig+14] should enable more uniform pulses. This is particularly important for improving the retrieval efficiency of the echo, which is crucial for the feasible operation of a spin ensemble as a quantum memory.

6.5 A multimode memory for coherent microwave pulses

A desired but feasible quantum memory would be able to store multiple photons. In a spin ensemble, the number of stored modes is limited by T_2/T_2^*, where $T_2^* = 1/\Gamma_2^\star$ is the total ensemble dephasing time. This can be very large in inhomogeneously broadened ensembles. In our experiments discussed in the previous chapters, $T_2/T_2^* \approx 266$. However, the finite bandwidth of the cavity stretches the emitted spin echo to 100 ns such that the effective maximum number of modes is reduced to approximately 56. Here, we demonstrate the potential of Er:YSO as a possible multimode quantum memory by storing and retrieving several weak coherent pulses. Storage of microwaves in the high power regime has been reported in Ref. [Wu+10]. A recent investigation by Grezes et al. [Gre+14] demonstrates storage of extremely weak microwave pulses (~ 3 photons) in an NV spin ensemble with an efficiency of 2×10^{-4}. In a multimode spin ensemble quantum memory, an incoming photon is mapped onto a coherent spin wave in the ensemble. For simplicity, we only consider the case of a single photon. The generalization to multi-photon and coherent states is straight forward as long as the number of excitations is much smaller than the number of spins in the ensemble. A single photon being absorbed by an ensemble of N spins manifests itself as a coherent superposition of all possibilities of one spin being excited with the rest in the ground state. In an inhomogeneously broadened ensemble, one has to consider the different precession frequencies of the spins, yielding

$$|\Psi\rangle_{\mathrm{ph}} = \frac{1}{\sqrt{N}} \sum_{k=1}^{N} |\downarrow_1 \downarrow_2 \cdots \uparrow_k \cdots \downarrow_N\rangle \, e^{-i\delta_k t}, \tag{6.5}$$

where δ_k denotes the detuning of the k-th spin from the mean precession frequency of the ensemble. As the time t progresses, this state dephases into a dark mode, such that the bright mode can absorb the next photon. The temporal separation between the application of storage pulses has to be larger than T_2^* [Wu+10], here $T_2^* \approx 22$ ns. Upon application of a π-pulse, the time evolution is reversed and all the dark modes rephase again emitting a photon. Thus, the time order of the incident photons is reversed with respect to the sequence of emitted photons.

Next, a proof of principle experiments is performed, which demonstrates the multimode storage capability of the Er:YSO spin ensemble by applying weak coherent MW pulses. All MW storage pulses have a length of 10 ns and their amplitude is controlled by the amplitude of the 200 MHz modulation pulse applied at the IF port of the mixer shown in Fig. 6.6. The pulse calibration can be found in appendix 9. Figure 6.15 shows a sequence of four irregularly spaced weak pulses, which are sent into the ensemble. A π-pulse situated in the middle of the plot reverses the time evolution of the ensemble. Note that the height of the π-pulse exceeds the vertical scale of the plot. After the refocusing pulse, all four stored pulses are retrieved in reversed time order. Each applied MW pulse has a power of approximately 25 μW. Since the pulse length is set to 10 ns, this corresponds to $\sim 10^{11}$ photons.

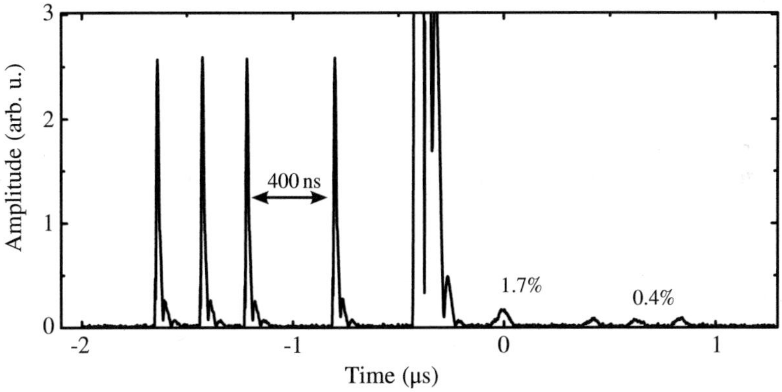

Figure 6.15: Storage and retrieval of four irregularly spaced MW pulses. The strong π-pulse in the middle extends far beyond the graph's vertical axis. The presented data is the result of 18 averages with a repetition rate of 10 mHz.

The storage efficiency for a given pulse is defined by the retrieved energy over the input energy. As one can see from Fig. 6.15, the recorded input pulses exceed the output pulses by far. Therefore, one can estimate the storage efficiency by comparing the respective recorded pulse energies. Experimentally, this is done by comparing the integral over the recorded power of the input pulse, with the integral over the recorded power of the output pulse. For this example, we determine an efficiency of 1.7% for the most recent pulse and approximately 0.4% for the rest.

In a second experiment, 16 evenly spaced weak pulses have been stored and retrieved. This time, the input pulses consist of $\sim 4.6 \times 10^{11}$ photons each. Figure 6.16 presents the complete recorded pulse sequence. Note that the detection of the weakest pulses required a large amplification of the signal. Therefore, the comparably strong input pulses operate in the saturation limit of our amplifiers. This makes them appear smaller than they actually are. In addition, the signal is averaged 595 times with a repetition rate of 10 mHz. As expected, the emitted echos decay towards longer storage times.

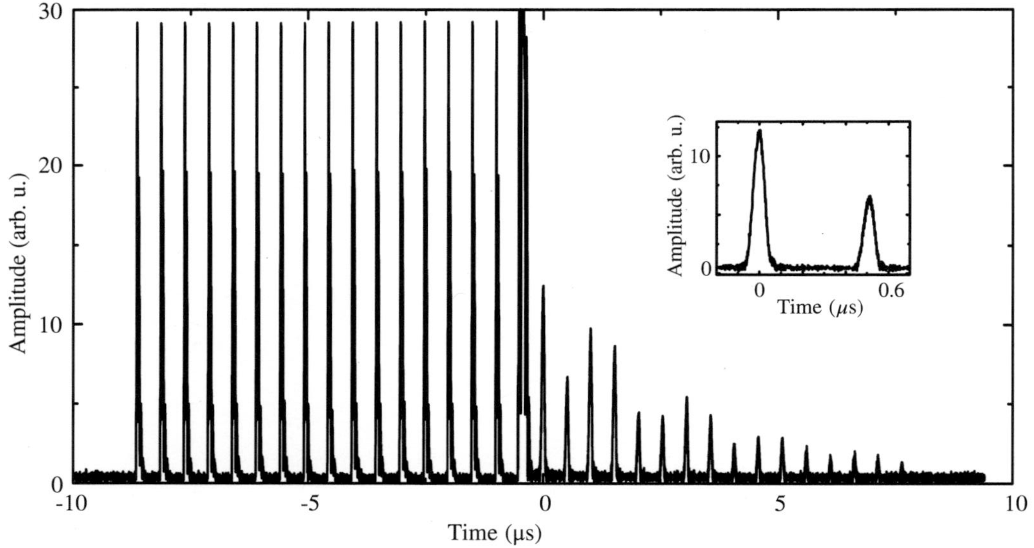

Figure 6.16: Amplitude vs. time of a 16 pulse sequence stored in the spin ensemble. All input pulses are of the same height and 10 ns width. Note that the strength of the input pulses is already in the saturation limit of our amplifiers, i.e. they appear smaller. The retrieved pulse stream shown in the plot is exponentially damped and exhibits oscillatory behavior. (Inset) Zoom on the first two pulses.

Interestingly, the decay pattern shows modulations attributed to electron spin echo envelope modulations (ESEEM), which have been investigated in Sec. 6.4. The performance of a memory is defined by the efficiency for a pulse being stored during the coherence time T_2 of the spin ensemble. For the present system, this corresponds to approximately 5.6 µs. To determine the overall memory performance, the pulse energy at 2.8 µs = $T_2/2$ after the refocusing pulse is analyzed and compared to the energy of the input pulse. Since the recorded input pulses operate in the saturation regime, we use the input energy values from the previous study (see Fig. 6.15) weighted with our calibration curve in appendix 9. The result is an energy retrieval efficiency of $\sim 2 \times 10^{-4}$, which is comparable to recent experiments with NV centers [Gre+14]. Another study working with classical pulses containing 10^{14} photons achieved an efficiency of 10^{-10} [Wu+10]. This shows that our setup is operating with a state of the art efficiency. Miniaturization of the resonator and the usage of parametric and cryogenic HEMT amplifiers may enable operation in the single photon regime.

According to Grezes et al. [Gre+14], the theoretical limitation for the efficiency is given by T_2. For the present values, this should give an efficiency of 10^{-3}. They point out, a non-Markovian bath can reduce the single spin Rabi oscillation time faster than the spin-echo decay time as observed in Ref. [HGA06]. The remaining order of magnitude may be attributed to the inhomogeneous AC field, which causes inaccurate π-pulses, such that only a fraction of the stored energy is refocused properly. The application of *optimal control* pulse schemes [Sig+14] would help to improve the yield. Furthermore, clever resonator designs with very homogeneous AC fields [Ben+13] in conjunction with surface spin doped samples [Pro+14a] would also contribute to a better efficiency.

However, significant progress can only be achieved with better T_2 times. Isotopically purified Er:YSO crystals reduce the amount of decoherence due to unwanted magnetic impurities. One order of magnitude is typically achieved by operating the memory at a *clock transition* where the spin is to first order insensitive to magnetic field fluctuations [Dol+11]. Specifically for spins in YSO, spectral diffusion due to the coupling to the Y^{89} nucleus and other electronic sub-ensembles might limit the overall performance. The theoretical analysis by Grezes et al. [Gre+14] shows that already an improvement in coherence by one order of magnitude would allow for efficiencies on the order of 0.1, enabling experiments in the quantum regime. This is certainly within reach.

7 A microwave-optical hybrid system

This chapter focuses on the optical properties of Er^{3+}:Y_2SiO_5. First, the optical transitions are investigated by continuous wave transmission spectroscopy. Then, the microwave-optical interaction of the Er^{3+}:Y_2SiO_5 hybrid system is studied. Specifically, the energy relaxation of the spin system is monitored optically. This is a first step towards the implementation of a coherent quantum converter.

7.1 Setup

Figure 7.1(a) shows a photograph of the sample holder used in the experiments. It has two slits which allow laser light to pass through the sample. The microwave part of the setup is identical to the one presented in Sec. 6. The laser is aligned such that it is as close as possible to the surface of the CPW resonator sketched in Fig. 7.1(b).

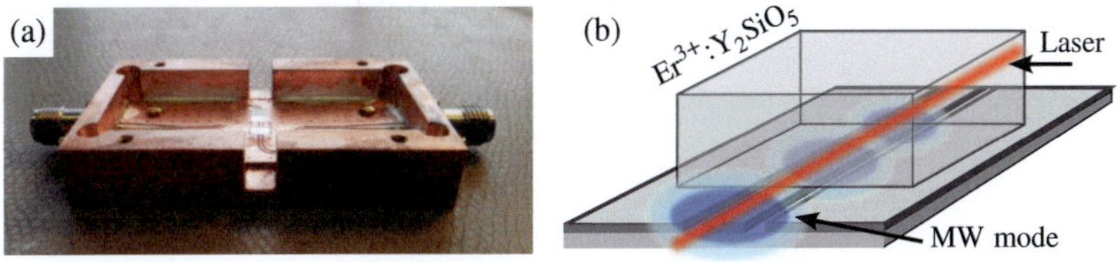

Figure 7.1: (a) Sample holder with microwave connectors on the left and right and slits in the front and back for the laser. In the middle, the Er:YSO crystal is seen. (b) Schematics of the hybrid system shown the coupling geometry where the optical beam (red) overlaps with the magnetic mode of the CPW resonator (blue).

Depending on the scope of the experiment, the optical spectroscopy is carried out either in pulsed mode at high powers (1 mW) or at low powers ($\sim$ 10 µW) and continuous wave. In both cases, we did not detect a rise in temperature, showing that the hybrid system is indeed operating at milli-Kelvin temperatures.

7.2 Optical spectroscopy of 50 ppm Er:YSO

First, the optical properties of the Er:YSO sample are investigated. The laser is tuned such that it sweeps continuously over the optical transition of a specific site at low power. The transmitted signal is recorded with a photo-detector. An optical transition shows up as a dip in the spectrum. By applying a magnetic field, the recorded dip splits up into maximum 8 dips. For every magnetic class *a* and *b* belonging to a crystallographic site, a 4-level structure is associated. If the corresponding optical ground and excited g-factors differ, all transitions can be seen.

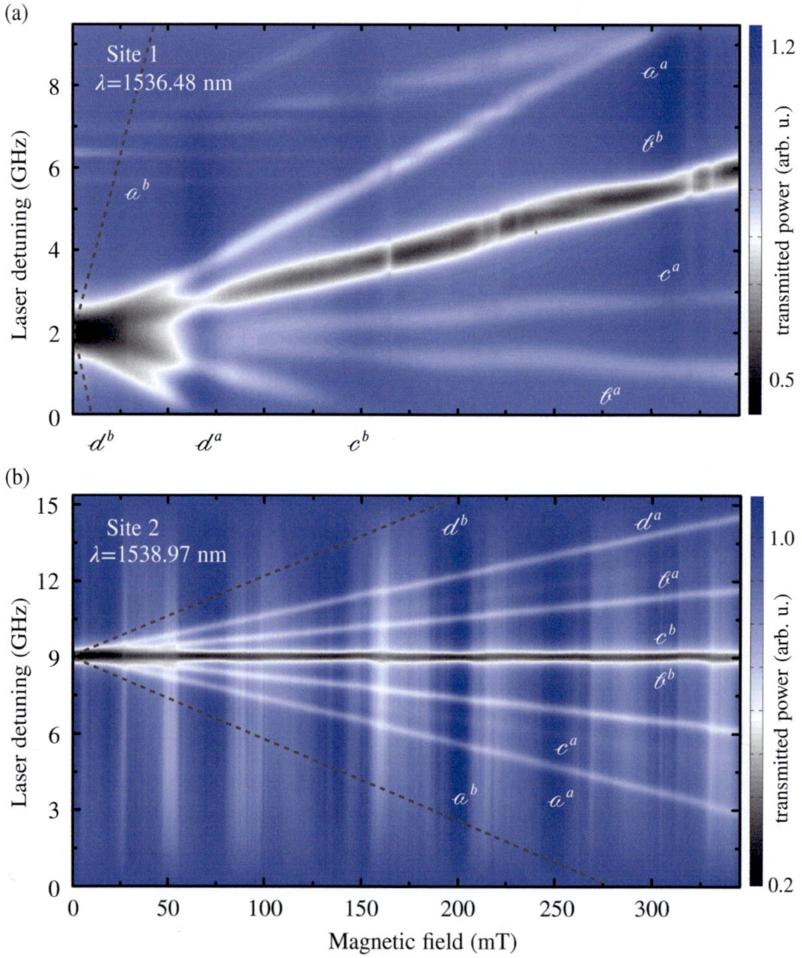

Figure 7.2: Optical transmission power spectrum vs. magnetic field of (a) site 1 and (b) site 2. The spectra were recorded at $\approx 10\,\mu\mathrm{W}$ laser power.

Figure 7.2 shows the so-called Zeeman spectrum for the optical sites 1 and 2. Here, the script characters $\mathscr{a}, \mathscr{b}, \mathscr{c}, \mathscr{d}$ indicate the optical transition and the superscript denotes the magnetic class [Sun+08]. For site 1, all transitions separate whereas at site 2, the $\mathscr{b}^b$ and $\mathscr{c}^b$ transitions coincide. The allowed transitions from the ground state ($\mathscr{b}$) appear strongest in the spectrum because the $|g\downarrow\rangle$ state is the most populated state at low temperatures. However, the recorded spectrum does not reflect the thermal distribution of the states due to optical pumping effects.

These would vanish at lower probing powers, but in this case, the larger power allows us to follow all absorption dips up to large Zeeman splittings. For the rest of the analysis, we focus on the transitions of site 2^a. These do couple to the high-field spin transition of the microwave resonator and allow us to study the microwave-optical interaction in the strong coupling regime.

At this point, it should be noted that in collaboration with D. Rieger, the optical coherence and T_1 times of the here studied sample have been determined employing optical two and three pulse echo schemes, respectively. The optical T_2 is approximately 28 µs and the T_1 time amounts for 1.5 ms [Rie13].

7.3 Optical detection of the spin relaxation

The experiment presented in this section attempts to monitor the population of the $|g\uparrow\rangle$ state after excitation with a microwave pulse by optically probing the c transition going from $|g\uparrow\rangle$ to $|e\uparrow\rangle$. This imposes some challenges: For instance, the HWHM inhomogeneous broadening of an isolated optical transitions at a typical magnetic field of 246 mT is $\sim$ 150 MHz, requiring broadband pulses to cover the full signal. This can be achieved by sweeping the frequency of the laser for the duration of the pulse. In addition, the pulse should be relatively short in order to obtain a sufficient time resolution. Furthermore, the relaxation time T_1 of the microwave transition requires a waiting time of $5T_1 \approx 100$ s in order to start off in a defined state. The optical setup has been modified accordingly and it is schematically displayed in Fig. 7.3. It is synchronized to the microwave time domain setup discussed in the previous chapter, which is used for the microwave pulses.

A global trigger initiates a fast frequency sweep of the laser and starts an arbitrary wave generator (AWG). The AWG generates a rectangular voltage pulse sent to the acousto-optical modulator, which translates it into an optical pulse. The timing is chosen such that this broadband pulse only covers the targeted optical transition. The signal is acquired in the usual way using photo detectors and an oscilloscope. Although the major part of the optical fibers are polarization maintaining fibers, the polarization slowly drifts on a timescale of minutes. In order to compare the absorption signals with confidence, a reference beam, which does not enter the cryostat, is recorded additionally. The analysis of the acquired absorption signal is performed by fitting a Gaussian line shape to the absorption dip. The second port of the global trigger unit starts the microwave pulse sequence. The corresponding setup is described in Fig. 6.6. The time delay between the two channels of the global trigger is tuned such that the microwave pulses always arrive first at the sample. A precise timing is not necessary because the optical setup probes the z-component of the spin system, i.e., its magnetization. From the spin lattice relaxation measurements, it is known that the longitudinal relaxation takes place on the time scale of seconds, which is very long compared to the experimental accuracy. Thus, although the coherence information decays fast, the z-component remains for a long time.

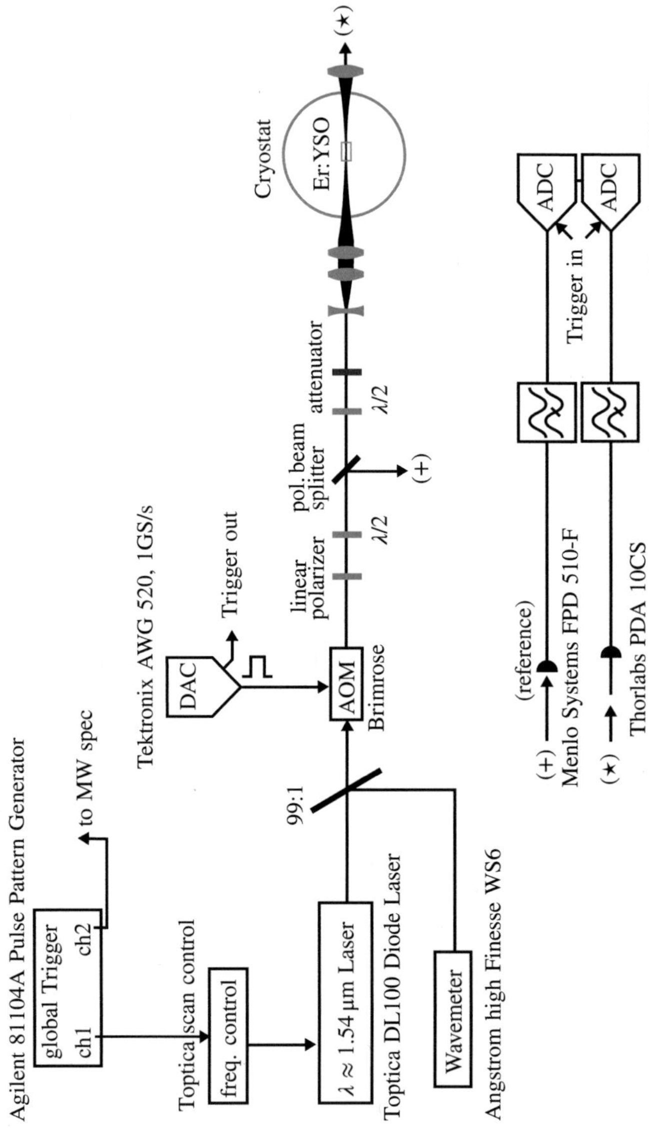

Figure 7.3: Optical broadband pulse generation and detection.

Figure 7.4(a) presents a saturation experiment where the c transition at 246 mT is probed optically while a continuous-wave microwave signal resonant with the spin transition is increased in power. The plot shows the depth of the absorption dip as a function of the applied microwave power. At −70 dBm, the absorption signal increases until it saturates at −40 dBm, indicating a saturation of the spin transition. We note that the temperature of the cryostat remained constant during the entire experiment.

Since the MW setup can supply short strong microwave pluses, it is possible to investigate the dynamics of the saturation process. In this experiment, short microwave pulses with fixed power and variable length are applied. The microwave power is chosen to 10 dBm at the sample as in the pulsed ESR experiments.

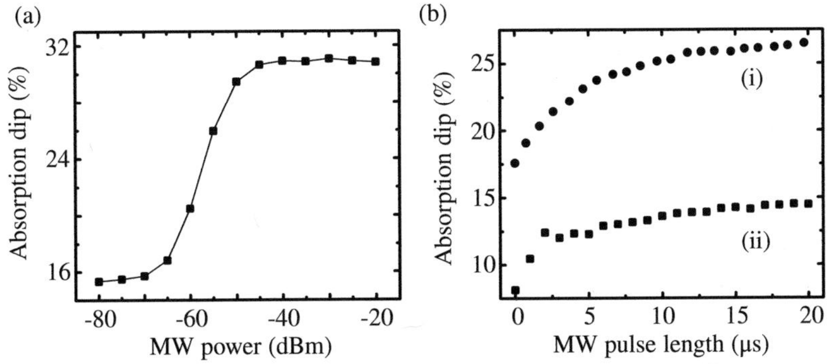

Figure 7.4: (a) MW power dependence of the optical c^a absorption dip. (b) Optical absorption signal of the c^a transition vs. MW pulse length. The power of the pulse is approximately 10 dBm at the crystal. Curve (i) was recorded using optical pumping to reset the system whereas curve (ii) was measured by waiting 100 s after each point to allow the spin ensemble to thermalize to the ground state (see text for details).

There are two methods to measure the spin relaxation. Graph (ii) in Fig. 7.4(b) has been recorded with the following pulse sequence. First, a resonant MW pulse excites the spin ensemble. Then, the c^a absorption dip is probed by a 200 μs long broadband optical pulse of 1 mW power. Before the next measurement is started, a waiting time of 100 s ensures that the spin system relaxes back to its ground state. This protocol has the obvious disadvantage that it takes a long time to record all the experimental points for different pulse lengths. This makes the experiment susceptible to low frequency noise produced by the intensity variations of the polarized laser beam. This results in a larger error bar of the experimental points. Here, a *reset button* would be preferable, which reinitializes the system in the ground state. This may be possible by employing higher levels in the optical spectrum.

A way to circumvent this problem is to achieve a partial reset by optical pumping. The essence of this procedure is to transfer a part of the population back to the ground state and erase all coherent information such that the system starts with a defined state, which can be any state with finite magnetization. The drawback is a reduced signal strength because a smaller fraction of the ensemble contributes to the signal. The procedure of acquiring the signal is identical to the procedure discussed before but instead of the 100 s waiting time, the following pumping scheme is applied. For a duration of 3 s, the c^a transition is probed with 200 µs long broadband pulses at a repetition rate of 50 Hz. The laser power is fixed to 1 mW. The pulses pump the population from the $|g \uparrow\rangle$ to the $|e \uparrow\rangle$ state. With a certain probability, some of the excitations will relax into $|g \downarrow\rangle$ state. The repetition rate is slow enough such that in between the pulses, the system has approximately 20 ms time to relax. This is enough time because the optical energy relaxation time has been determined to be approximately 1.5 ms [Rie13]. Graph (i) in Fig. 7.4(b) shows the result, which has been recorded in a fraction of the time compared to Graph (i). The optical absorption data shows a nice and smooth dependence on the length of the input MW pulse. The equilibrium depth resides on a higher level, indicating that the applied pumping scheme does not establish full polarization, which is expected. Both measurements clearly demonstrate the observation of the spin dynamics using an optical readout. One should point out that thermal effects cannot be responsible for the observed phenomena, because the spin-lattice coupling is weak with relaxation rates in the sub-Hz regime.

Although the applied microwave pulses are in the correct regime of coherent manipulation, no Rabi oscillations are detected. Note that a pulse length of about 40 ns at the current power corresponds to an effective π rotation of the spins detected by the spin echo experiments in the previous chapter. The absence of Rabi oscillations in the measurement originates from the inhomogeneous magnetic AC field of the CPW resonator. This field decays in a first approximation inversely to the distance from the central conductor of the resonator, giving rise to a continuous distribution of Rabi frequencies. The optical probe beam only measures an ensemble average inhibiting the detection of an oscillatory signal. Such challenges may be addressed by optimal control pulse schemes [Sig+14].

These findings now allow to move a step further. Figure 7.5(a) shows a cut of the optical Zeeman spectrum of Fig. 7.2(b) at 246 mT, which is the magnetic field at which the S_{2a} transition is in resonance with the microwave resonator. From the cw optical spectroscopy, all transitions can be identified and the corresponding absorption dips are labeled accordingly in the figure.

In the following experiment, the spin ensemble is saturated by a 3 s long MW pulse with a power of approximately 10 µW at the sample. The laser power is set to 100 µW. After the MW pulse is switched off, the laser probes the absorption dip of the c^a transition with the broadband pulse technique described above. Every second, such a short pulse is sent to monitor the time evolution of the depth of the absorption dip. Specifically, the absorption dip is fitted by a Gaussian function and the depth is normalized to the baseline of the off-resonant transmission. Figure 7.5(b) shows the resulting time dependence of the c^a absorption. Since the probing power of the laser is comparably weak, the decay curve reflects the decay of the $|g \uparrow\rangle$ state. An exponential fit yields a T_1 time of 20.8 ± 1 s. For consistency, the d^a transition is probed. This transition connects the $|g \uparrow\rangle$ state with the $|e \downarrow\rangle$ state. Note, that the labels of the state vectors imply that this transition does not conserve the angular spin momentum and should be forbidden. However, the presence of the crystal field effects a mixing of the eigenstates compared to the free ion such that the usual selection rules do not apply. Nevertheless, the transition shows a slightly smaller absorption signal. Figure 7.5(b) presents the corresponding decay curve. One can see that the decay is slightly faster, caused by optical pumping of the probe laser. Population, which is transferred to the $|e\downarrow\rangle$ state, is more likely to decay into the ground state $|g\downarrow\rangle$. This accelerates the measured decay.

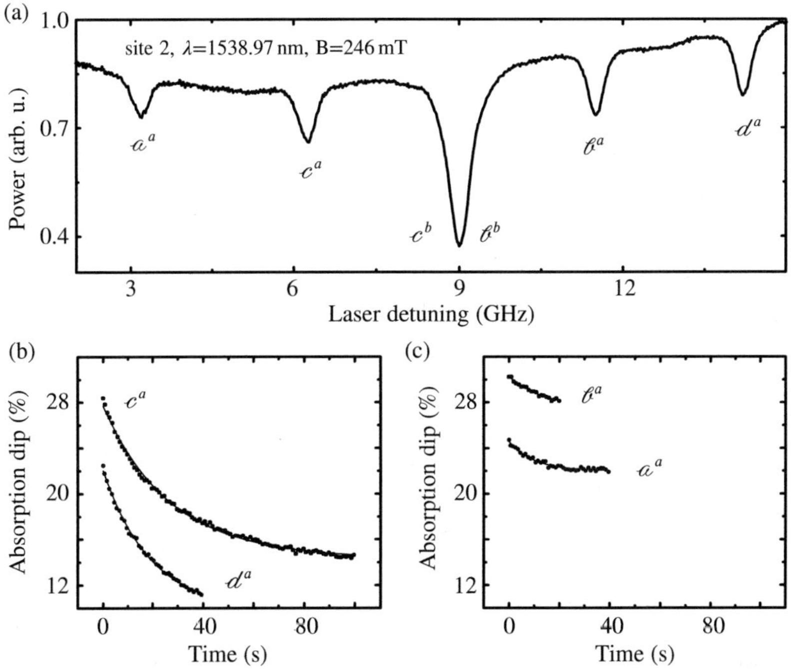

Figure 7.5: (a) Optical transmission at 246 mT. (b) Time evolution of the absorption dips of transitions c^a and d^a after a strong microwave pulse on the $|g\downarrow\rangle$ to $|g\uparrow\rangle$ transition of site 2a (site 2^a). (c) Corresponding time evolution of the absorption dips of the a^a and b^a transitions.

To complete the analysis, we probe the two transitions originating from the $|g\downarrow\rangle$ state. The result is displayed in Fig. 7.5(b). Both absorption signals from the a^a and b^a transitions weaken slightly over time. In fact, one would have anticipated that the absorption dips grow as the spins relax to the ground state. Here, the signal is entirely masked by optical pumping effects. The stroboscopic laser beam pumps population into the $|g\uparrow\rangle$ state. Unfortunately, this effect dominates the spin relaxation signal. Moreover, the laser beams probing the $|g\uparrow\rangle$ state address a much larger volume because outside the copper cavity's effectiv mode volume, most of the spins are in the ground state. The present experimental setup did not allow a study at lower laser probe power or techniques involving averaging. The latter is inhibited by the low noise frequency power fluctuations of the laser, which destroy the small signal. Despite these issues, the experiment gives an impression of the complexity of building a coherent microwave to optical converter. The measured T_1 relaxation time is in agreement with its determination by microwave saturation spectroscopy, see Sec. 6.1. In that respect, these experiments can be understood as a proof of concept to see what is possible. In future experimental investigations, optical pumping combined with specifically engineered MW pulses may eventually enable coherent transfer of quantum information. A theoretical proposal is presented in Sec. 3.4. A crucial point for the practical implementation is the possibility of a quick reset of the system to its ground state. This may be achieved by optical pumping to higher levels of the Er^{3+} ion, requiring lasers with shorter wavelengths [Dou+95].

8 Conclusions

In this thesis, the microwave and optical properties of rare earth spin ensembles have been investigated with particular focus on the rare-earth (RE) element erbium. The goal of the thesis was to explore the potential of RE based hybrid quantum systems for applications in quantum computing and quantum communication.

The first part of the thesis deals with the coupling of rare-earth spin ensembles to superconducting (SC) lumped element (LE) resonators. We identified all four erbium sub-ensembles in the low temperature on-chip microwave spectroscopy. Subsequently, the angular magnetic anisotropy of the Er:YSO crystals has been analyzed. For specific orientations of the static magnetic field, the ensemble possesses large effective g-Factors, which result in a stronger Zeeman effect. This lowers the magnetic field requirements for tuning the spins into resonance with the SC circuit. We found, however, that in order to enter the strong coupling regime, one has to scarify the large spin tuning rate of Er:YSO: At high DC g-factors the coupling strength is small and the inhomogeneous spin linewidth is high, whereas at low DC g-factors the coupling strength is larger and the linewidth is small. By probing this anisotropic behavior, we have been able to determine the correct parameter regime for achieving strong coupling between the resonator and the spin ensemble. This means, the collective coupling strength dominates both the dissipation rate of the resonator and the inhomogeneous spin linewidth of the ensemble. A collective coupling strength of $\upsilon_D/2\pi = 34\,\mathrm{MHz}$ and an inhomogeneous linewidth of $\Gamma_2^\star/2\pi = 12\,\mathrm{MHz}$ has been determined. These findings were published in Ref. [Pro+13]. The temperature dependencies confirmed the paramagnetic properties of the investigated spin ensembles and the spin lattice relaxation time has been determined to 4.3 s for the 200 ppm Er:YSO sample.

Having demonstrated strong coupling of an erbium spin ensembles to a SC LE microwave resonator, issues regarding the practical integration of RE spin ensembles with SC quantum information circuits were addressed. Employing $YAlO_3$ (YAlO) as a new host material with a higher crystal symmetry, we managed to decrease the magnetic field requirements for strong coupling from 250 mT to 36 mT [Tka+14]. In a further study, we investigated SC circuits fabricated directly onto a Nd:YAlO substrate and we confirmed strong coupling, too. A third step towards a feasible application of rare earth spin ensembles is the local integration of ions into the crystal. This allows one to realize local memory elements in a SC circuit which do not interfere with the rest of the circuitry. Together with our collaborators at RUB Bochum, focused ion beam implanted Er:YSO crystals were investigated [Kuk+14]. We probed the implanted ensembles by on-chip ESR spectroscopy and studied the dependence of the coupling strength with respect to the number of implanted ions. It has not been possible to enter the strong coupling regime due to a larger inhomogeneous spin linewidth compared to bulk-doped crystals. We believe that the linewidth could be decreased by hot implantation and an improved annealing procedure. The temperature dependence confirmed the paramagnetic properties of the ensemble. These experiments demonstrate the high potential of this approach [Pro+14a].

Since the first part of the theses has been devoted to the study of the magnetic anisotropy of the Er:YSO spin ensemble, it is interesting to investigate the AC g-factor anisotropy as well. However, this is not possible in the planar circuit architecture, and therefore, a 3D cavity has been employed. Here, a distinct electromagnetic mode ensured a homogeneous AC field at the position of the crystal. By rotating the crystal with respect to the magnetic field, we determine a change of the AC g-factor by a factor five. At maximum AC g-factor, the 3D cavity showed strong coupling to the Er:YSO spin ensemble [Pro+14c].

To complete the characterization of rare earth based hybrid systems, we developed a new setup for time domain investigations of a 50 ppm doped Er:YSO crystal. Here, a non-superconducting coplanar waveguide copper resonator with a relatively large mode volume was employed allowing for simultaneous microwave and optical spectroscopy due to the overlap of optical and microwave modes. This configuration served as a test-bed to explore the crucial elements of a future quantum memory and quantum converter. Since the setup is different to the previous investigations, we first characterized it by continuous wave microwave spectroscopy. As a result, all erbium sub-ensembles were detected and despite the fact that the resonator featured a very low g-factor, the high-field transition is strongly coupled. The hybrid system operated in an interesting regime, where the inhomogeneous spin linewidth and the homogeneous linewidth of the copper resonator were on the same order of magnitude. Spin saturation spectroscopy revealed a T_1 of approximately 20 s at 30 mK.

By applying magnetic field gradients, we were able to manipulate the inhomogeneous spin linewidth. As the inhomogeneity is increased, the hybrid system underwent a transition from the strongly coupled to the weakly coupled regime. Interestingly, the power transmission spectra showed unexpected shapes, which were not covered by the theory of two coupled quantum harmonic oscillators. At the border from the strong to the weak coupling regime, the mode splitting transformed into a three peak structure followed by a single resonance peak in the weak coupling regime. Moreover, the data suggested an increase in coupling strength by a factor of 1.5 while the inhomogeneous spin linewidth increases by a factor of 3. Up to now, the details of the observed phenomena are subject to theoretical investigations and may originate from the interplay of the *homogeneously* broadened harmonic oscillator and the *inhomogeneously* broadened effective spin oscillator system.

Next, we probed the same spin ensemble by short microwave pulses. Specifically, the strongly coupled high-field transition has been in the center of the investigation. As a response to a 20 ns long pulse, the system showed coherent vacuum Rabi oscillations between the hybridized modes of the spin ensemble and the copper cavity. They match the coupling strength determined by continuous wave spectroscopy. In order to probe the coherence time T_2 of the Er:YSO spin system, we used a two pulse echo sequence. The echo decay curve revealed a T_2 of 5.6 µs. Interestingly, we observed prominent oscillatory modulations of the echo decay. Those are attributed to super-hyperfine dipole-dipole interactions with the nuclear spin of the surrounding $^{89}Y^{3+}$ ions. To achieve a better frequency resolution, the so-called electron spin echo envelope modulations (ESEEM) were probed by a stimulated echo sequence, which has a longer decay time. The oscillations matched the expected frequency range around the free precession frequency of the $^{89}Y^{3+}$ nuclear spin of approximately 500 kHz. Based on the crystal Er:YSO structure, we have performed a numerical simulation of the ESEEM spectrum which also showed good agreement with the measurement.

As the temperature was increased, the decoherence rate grew. We found that this temperature dependent decoherence arises predominantly from spin flip-flop processes of the other three erbium sub-ensembles present in the crystal. At temperatures below 50 mK, T_2 remained constant indicating that those fluctuations are frozen out. The residual decoherence has two dominant contributions. At 30 mK, spin flips of the non-polarized nuclear spins of the ^{167}Er isotope contribute as well as dipole-dipole interactions withing the same spin ensemble. In order to probe the magnitude of the latter contribution, we employed a modified two pulse echo sequence where the magnitude of the refocusing pulse was varied. Despite the challenges imposed by the inhomogeneous AC probe magnetic field of the resonator, a dipole-dipole coupling strength of $\upsilon_D/2\pi \approx 11.9$ kHz was extracted. This sets a theoretical limit of ~ 13.4 µs on the maximum T_2 time achievable using the present Er:YSO crystal with a concentration of 50 ppm. Since the dipole-dipole interaction strength scales quadratically with the DC g-factor, spin ensembles with large g-factors have a two orders of magnitude larger υ_D, which contributes significantly to the linewidth.

Next, we tested a prototype multimode quantum memory scheme. Four irregularly spaced low power coherent microwave pulses were stored in the Er:YSO spin ensemble and retrieved. Each coherent pulse consisted of approximately 10^{11} photons. We even managed to store and retrieve up to 16 pulses containing 4.6×10^{11} photons each. The retrieval efficiency was determined to be 10^{-4}. Although this appears to be rather low, it is on the same level as state of the art experiments with NV centers [Gre+14]. In fact, most of the efficiency depends on the magnitude of T_2 and only one order of magnitude is lost due to imperfect refocusing pulses and related experimental details. Improving T_2 by one order of magnitude would already boost the efficiency significantly. In the case of erbium spin ensembles, a lower concentration of the spins as well as isotopically purified samples are necessary. Regarding the constraints imposed by the dipole-dipole coupling, a concentration of 5 ppm Er:YSO, would increase the maximum achievable coherence time above 100 µs. Also, it would be beneficial to position the crystals such that their magnetic classes overlap forming a single spin ensemble.

Finally, we investigated the hybrid system by optical spectroscopy at 1.54 µm. By applying a magnetic field, the full Zeeman spectrum was measured and all transitions were identified. Then, the microwave-optical interaction was studied by probing the optical transitions and applying microwave pulses. In order to achieve a reasonable signal-to-noise ratio and cover the full bandwidth of the optical transition, we developed a setup generating short broadband optical pulses. This enabled us to record the energy relaxation of the spin system by stroboscopically probing one of the optical transitions. This may be the first step towards the implementation of a coherent quantum converter.

Outlook

In the past years, hybrid quantum systems received a lot of attention. The coherent interaction between quantum systems of different physical nature is interesting on both a fundamental level and with regard to applications. In this theses, we have coherently interfaced solid state spin ensembles with SC quantum circuits. Although spin systems have been investigated for decades with electron spin resonance using 3D cavities, the combination with the emerging field of circuit QED has created a research field of its own. The strong coupling of those coherent microwave circuits to solid state spin ensembles has shed new light on those systems. For instance, the physics of the collective coupling to inhomogeneously broadened spin ensembles is not yet fully understood. Specifically, our experiments on the low quality factor resonator yielded interesting spectra and there is ongoing theoretical research activity investigating the influence of the spectral shape of the inhomogeneous spin linewidth on the coupling physics. In this context, it has already been shown that distinct spectral shapes give rise to new effects such as *cavity protection* where the dephasing into dark modes is suppressed.

The results on the focused ion beam implantation of erbium ions into Y_2SiO_5 substrates are also interesting from a material science perspective. In order to reduce the inhomogeneous spin linewidth, a more detailed understanding of the process of erbium ions entering the substrate and substituting the yttrium ions is required. In order to improve the yield of this mechanism, current efforts rely on implantation in hot substrates. Once focused ion beam implantation of erbium enables high quality strongly coupled spin ensembles, it will be a key technology towards the practical integration of spins in circuit QED processors. This will also allow the integration of on-chip optical waveguides with erbium implanted substrates [SPT13], which may lead to combined on-chip microwave-optical spectroscopy and a prospective on-chip quantum media converter.

An exciting question is: *Is it possible to build a rare earth based single photon and multimode quantum memory?* The experiments in this thesis demonstrate the multimode storage capability of erbium doped Y_2SiO_5. Although the setup is not optimized, it achieves a good fidelity. The extension of the presented pulse sequences to the few photon coherent state regime is straightforward. In order to achieve the necessary signal-to-noise ratio, a Josephson parametric amplifier in series with a cryogenic HEMT amplifier has to be employed. Furthermore, a smaller superconducting cavity is required, in order to reduce the power requirements for the π-pulse. The high power pulses used in the present experiments would destroy a cryogenic HEMT amplifier. To boost the efficiency, adiabatic optimal control pulses [Sig+14] have to be applied for achieving an accurate refocusing. Next, lower ion concentrations will provide an increase in coherence due to the reduction of dipole-dipole interactions. With these improvements, few to single photon fidelities of 0.1 may be in reach within the next years [Gre+14]. In parallel, the operation at lower magnetic fields based on Er^{3+}:$YAlO_3$ crystals and the incorporation of qubits in the circuit is realistic on a short timescale. However, the presence of the Al nuclear spin in $YAlO_3$ may hinder a larger coherence. On the other hand, the decoherence may be suppressed by a frozen core effect as in the case of Er^{3+}:Y_2SiO_5. Based on the experience with hybrid circuit QED systems acquired in the last years, a proof of concept multimode single photon quantum memory is certainly realizable within the next three to five years.

Another exciting yet more challenging question is: *Is it possible to construct an erbium based microwave to optical quantum converter?* The combination of optical and microwave spectroscopy at milli-Kelvin temperatures opens up a whole new set of challenges. In this thesis, a basic setup has been investigated illuminating the complexity of this task. The planar architecture of the on-chip microwave resonators introduces field inhomogeneities, which require the application of adiabatic optimal control pulse schemes. An interesting way to circumvent this is to use a 3D microwave cavity. The larger geometric dimensions allow for more flexibility. This allows easy optical access and homogeneous microwave magnetic fields at the same time. However, the core problem of the large inhomogeneous broadening of the optical transition remains. Recently, a theoretical scheme for coherent optical to microwave conversion of single photons has been proposed [OBr+14]. Despite the technical challenges, reversible coherent microwave to optical conversion of quantum states may be possible. The question is, whether this will ever be feasible with a reasonable fidelity. From today's perspective, architectures based on nanomechanical resonators operating in the single photon regime may be easier to implement in the short-term because they rely on a less complex conversion scheme [And+14].

The hybrid system Er:YSO or similar rare earth compounds coupled microwave circuits may be of use in the research on quantum metamaterials. There, coupled multi-qubit and multi-resonator systems are investigated to manipulate the propagation of light in an engineered way. A simple metamaterial consisting of many qubits coupled to a common microwave waveguide [Zag+09] is easily realized by coupling an Er:YSO crystal to a long microwave transmission line. On a fundamental level, combined multi-wavelength single photon spectroscopy of Er:YSO or related compounds may enable the investigation of coherent quantum energy transport in multi-scale solid state systems.

In any scenario, hybrid systems based on rare earth ions doped into crystals face a bright and diverse future.

9 Zusammenfassung

In dieser Arbeit wurden optische Eigenschaften und Mikrowelleneigenschaften eines Seltenerdspinensembles untersucht, wobei im Zentrum das Seltenerdelement Erbium stand. Ziel war dabei die Untersuchung des Anwendungspotentials seltenerdbasierter hybrider Quantensysteme im Hinblick auf deren Integration in Quantencomputern und für die Quantenkommunikation.

Im ersten Teil wurde die Kopplung der Seltenerdspinensemble an supraleitenden (SL) *lumped element* (LE) Resonatoren untersucht. Alle vier Teilensembles des Erbiums konnten durch *on-chip* Tieftemperatur-Mikrowellenspektroskopie identifiziert werden. Anschließend wurde die magnetische Anisotropie des Er:YSO Kristalls winkelaufgelöst untersucht. Unter gewissen Magnetfeldorientierungen ergeben sich hohe DC g-Faktoren, die zu einem verstärkten Zeeman Effekt führen. Dadurch werden geringere Magnetfelder benötigt um die Spins in Resonanz mit dem SL Schaltkreis zu bringen. Es stellte sich jedoch heraus, dass zur Erreichung des starken Kopplungsregimes auf diesen Vorteil verzichtet werden muss: Bei hohen DC g-Faktoren ist die Kopplungsstärke klein und die inhomogene Linienverbreiterung des Spinensembles groß. Umgekehrt ist bei kleinen DC g-Faktoren die Kopplung größer und die Linienbreite kleiner. Durch systematische Analyse dieser Anisotropie konnten die nötigen Parameter für das starke Kopplungsregime ermittelt und starke Kopplung nachgewiesen werden. In diesem Regime dominiert die Kopplungsstärke die Dissipationsrate des Resonators und die inhomogene Spinlinienbreite des Ensembles. Die gemessene kollektive Kopplungsstärke betrug $v_D/2\pi = 34\,\mathrm{MHz}$ bei einer inhomogenen Spinlinienbreite von $\Gamma_2^\star = 12\,\mathrm{MHz}$. Diese Ergebnisse wurden in [Pro+13] veröffentlicht. Die Temperaturabhängigkeiten der untersuchten Spinensemble bestätigten deren paramagnetische Eigenschaften. Die Messung der Spin-Gitter-Relaxationszeit in einem 200 ppm dotierten Er:YSO Kristall ergab 4, 3 s.

Mit dem Nachweis starker Kopplung der Erbium Spinensemble an LE Mikrowellenresonatoren wurde ein wichtiger Meilenstein erreicht. Dies erlaubte den nächsten Schritt, der sich mit der Frage nach der praktischen Integration von Seltenerdspinensemblen beschäftigte. Hier wurde $YAlO_3$ (YAlO) als neuer Wirtskristall eingesetzt. Die höhere Kristallsymmetrie ermöglichte eine deutliche Reduzierung der notwendigen Magnetfeldstärke von 250 mT auf 36 mT [Tka+14]. In einem weiteren Experiment wurde die Kopplung von SL Schaltkreisen untersucht, die direkt auf einem Nd:YAlO Kristall hergestellt wurden. Hierbei wurde ebenfalls das starke Kopplungsregime erreicht. Ein weiterer Aspekt, der für die praktische Anwendung von Seltenerdspinensemblen in SL Quantenprozessoren essentiell ist, liegt in der ortsaufgelöste Integration von Seltenerdionen in einen Kristall. Dadurch ist es möglich lokale seltenerdbasierte Speicherelemente zu definieren, ohne den Rest des Schaltkreises zu stören. In Zusammenarbeit mit unseren Projektpartnern von der RUB Bochum sind YSO Kristalle untersucht worden, die mittels fokussierter Ionenimplantation (FIB) mit Erbium dotiert wurden [Kuk+14]. Die implantierten Ensemble konnten ebenfalls durch *on-chip* ESR Spektroskopie detektiert werden. Außerdem wurde die Abhängigkeit der Kopplungsstärke von der implantierten Konzentration untersucht. Hier wurde das starke Kopplungsregime nicht erreicht, da, verglichen mit den volumendotierten Kristallen aus den vorangegangenen Experimenten, die inhomogenen Linienbreite des Spinensembles erhöht war. Wir glauben, dass diese Linienbreite durch heiße Implantation und ein verbessertes Ausheilprotokoll weiter reduziert werden kann. Die Temperaturabhängigkeit bestätigte die paramagnetischen Eigenschaften des Ensembles. Alles in allem demonstrieren die gezeigten Experimente das große Potential dieses Ansatzes [Pro+14a].

In einer zweidimensionalen Resonatorarchitektur, wie sie in den vorangegangenen Experimenten verwendet wurde, ist die Untersuchung des AC g-Faktors nicht ohne weiteres möglich. Daher wurde ein 3D Resonator verwendet, dessen spezielle elektromagnetische Mode für ein räumlich homogenes AC Magnetfeld am Ort des Kristalls sorgt. Durch Rotation des Kristalls bezüglich des externen statischen Magnetfelds änderte sich der AC g-Faktor um eine halbe Größenordnung. Bei maximalem g-Faktor zeigte der 3D Resonator starke Kopplung an das Er:YSO Spinensemble [Pro+14c].

Zur weiteren Charakterisierung eines 50 ppm dotierten Er:YSO Kristalls wurde ein neuer Aufbau entwickelt, der eine zeitaufgelöste Untersuchung ermöglichte. Hier wurde ein nichtsupraleitender koplanarer Kupferresonator mit einem vergleichsweise großen Modenvolumen eingesetzt. Dieser ermöglichte gleichzeitige optische und Mikrowellenspektroskopie aufgrund des Überlapps beider Moden. Diese Anordnung diente als Testumgebung, um die wesentlichen Elemente zukünftiger Quantenspeicher und Quantenkonverter zu untersuchen. Im Vergleich zu den vorangegangen Untersuchungen wies der Aufbau einige Unterschiede auf. Daher wurde er zunächst mit Dauerstrichspektroskopie untersucht. Alle Erbium Teilensembles konnten detektiert werden, und trotz des niedrigen Gütefaktors des Resonators war der Hochfeldübergang stark gekoppelt. Das Hybridsystem arbeitete in einem bislang wenig untersuchten Bereich mit inhomogener Spinlinienbreite und homogene Linienbreite des Kupferresonators von gleicher Größenordnung. Mittels Spin-Sättigungsspektroskopie wurde eine T_1 Zeit von circa 20 s bei 30 mK ermittelt.

Durch Anlegen von magnetischen Feldgradienten war es möglich, die inhomogene Linienbreite des Spinensembles zu beeinflussen. Mit wachsender Inhomogenität durchlief das Hybridsystem den Übergang vom starken hin zum schwachen Kopplungsregime und interessanterweise zeigten die Leistungsspektren unerwartete Formen, die nicht mit der Theorie zweier gekoppelter quantenmechanischer harmonischer Oszillatoren erklärt werden konnten. An der Grenze vom starkem zum schwachem Kopplungsregime transformierte sich die Modenaufspaltung zu einer Struktur bestehend aus drei Peaks, die anschließend zu einem einzigen Peak im schwachen Kopplungsregime übergingen. Darüber hinaus suggerierten die Daten einen Anstieg der Kopplungsstärke um einen Faktor $1,5$, während die inhomogene Linienbreite gleichzeitig um einen Faktor 3 anwuchs. Derzeit sind die Details der beobachteten Phänomene Gegenstand theoretischer Untersuchungen. Möglicherweise kommen sie durch das Wechselspiel eines *homogen* verbreiterten harmonischen Oszillators und eines *inhomogen* verbreiterten effektiven Spinoszillators zustande.

Im Weiteren wurde das Spinensemble mit kurzen Mikrowellenpulsen untersucht. Der stark gekoppelte Hochfeldübergang stand dabei im Zentrum der Untersuchungen. Als Antwort auf einen 20 ns langen Puls zeigte das Hybridsystem kohärente Vakuum-Rabi-Oszillationen zwischen den hybridisierten Moden von Spinensemble und Kupferresonator. Diese stehen im Einklang mit der Kopplungsstärke, die über die vorangegangen Dauerstrichspektroskopie ermittelt wurde. Um die Kohärenzzeit T_2 zu bestimmen, wurde eine Zweipuls-Echosequenz verwendet. Die Zerfallskurve ergab ein T_2 von $5,6\,\mu s$. Interessanterweise war die Zerfallskurve von deutlichen Oszillationen geprägt. Diese sind auf eine Superhyperfeinwechselwirkung mit dem Kernspin der umgebenden $^{89}Y^{3+}$ Ionen zurückzuführen. Um eine bessere Frequenzauflösung zu erreichen, wurde eine stimulierte Echosequenz verwendet, die eine größere Zerfallszeit aufweist. Die gemessenen Oszillationen entsprachen dem erwarteten Frequenzregime nahe der freien Präzisionsfrequenz des $^{89}Y^{3+}$ Kernspins von circa 500 kHz. Dies Spektrum konnten wir durch eine numerische Simulation auf Basis der Er:YSO Kristallstruktur bestätigen.

Bei steigender Temperatur wuchs die Dekohärenzrate und die Analyse zeigte, dass diese Temperaturabhängigkeit im Wesentlichen auf Spin flip-flop Prozesse der anderen drei Teilensemble zurückzuführen ist. Unterhalb von 50 mK blieb T_2 konstant, da dort die magnetischen Fluktuationen der anderen Ensemble ausgefroren sind. Die verbleibende Dekohärenzrate hat zwei dominante Beiträge. Bei 30 mK ist der Kernspin des ^{167}Er Isotops noch nicht vollständig polarisiert. Dazu kommen Dipol-Dipol Wechselwirkungen innerhalb des selben Teilensembles. Um die Größe des letzteren Beitrags abzuschätzen, wurde eine modifizierte Spinecho-Sequenz verwendet, bei der die Größe des Refokussierungspulses variiert wurde. Trotz der Herausforderungen durch das inhomogene AC Magnetfeld, konnte eine Dipol-Dipol Kopplungsstärke von $v_D/2\pi \approx 11,9\,\text{kHz}$ ermittelt werden. Für den aktuellen Er:YSO Kristall mit einer Ionenkonzentration von 50 ppm führt dies zu einer theoretischen Grenze für T_2 von $\sim 13,4\,\mu s$. Da die Dipol-Dipol Wechselwirkung quadratisch mit dem DC g-Faktor skaliert, ist v_D in Spinensemblen mit großem g-Faktor bis zu zwei Größenordnungen größer und trägt damit signifikant zur Linienbreite bei.

Zudem wurde ein Prototyp eines Quantenspeicherprotokolls getestet. Vier unregelmäßig angeordnete, kohärente Pulse niedriger Leistung wurden im Er:YSO Spinensemble gespeichert und anschließend ausgelesen. Jeder dieser kohärenten Pulse entsprach 10^{11} Photonen. Es war uns sogar möglich bis zu 16 Pulse mit je $4,6 \times 10^{11}$ Photonen zu speichern und auszulesen. Die Effizienz des Speichers war 10^{-4}. Diese Zahl darf nicht unterschätzt werden, denn im Vergleich mit Experimenten an NV-Zentren in Diamant [Gre+14] erreichte der Speicher der aktuellen Untersuchung die gleiche Effizienz. Tatsächlich wird die Effizienz im Wesentlichen durch die Größe der T_2 Zeit bestimmt. Die Analyse zeigte, dass nur etwa ein Faktor 10 durch unvollkommene Refokussierungspulse und verwandte experimentelle Details verloren wurde. Eine Verbesserung der T_2 Zeit von einer Größenordung hingegen, kann die Effizienz des Speichers erheblich steigern. Im Falle von Erbium Spinensembles sind dafür eine niedrigere Spinkonzentration sowie isotopenreine Proben nötig. Die durch die Dipol-Dipol Wechselwirkung limitierte T_2 Zeit würde im Falle einer Konzentration von 5 ppm bereits auf jenseits von $100\,\mu s$ steigen. Ebenfalls wäre es von Vorteil die Kristalle so auszurichten, dass ihre magnetischen Klassen überlappen und sie damit ein einziges Spinensemble bilden.

Abschließend wurde das selbe Hybridsystem optisch bei $1,54\,\mu m$ spektroskopiert. Durch Anlegen eines Magnetfeldes wurde das gesamte sogenannte Zeemanspektrum vermessen, bei dem alle optischen Übergänge zugeordnet werden konnten. Danach wurde die Interaktion zwischen optischen und Mikrowellenübergängen untersucht, indem die optischen Übergänge vermessen wurden, während ein Mikrowellenpuls einen Spinübergang trieb. Um ein akzeptables Signal-zu-Rauschverhältnis zu erreichen, wurde ein Aufbau entwickelt, der kurze, breitbandige Pulse erzeugen konnte. Dieser erlaubte die Detektion der Spinrelaxationszeit durch stroboskopische Messung eines bestimmten optisch Übergangs. Dies ist ein erster Schritt in die Richtung der Entwicklung eines kohärenten Quantenkonverters.

Ausblick

Hybride Quantensysteme haben in den letzten Jahren viel Beachtung erfahren. Die kohärente Wechselwirkung zwischen unterschiedlichen physikalischen Quantensystemen ist interessant für fundamentale Fragestellungen, wie auch für Anwendungen. Diese Arbeit hat sich mit der Kopplung von Festkörperspinensembles mit SL Quantenschaltkreisen beschäftigt. Auch wenn Spinsysteme schon seit Jahrzehnten mit Elektronspinresonanz in 3D Resonatoren untersucht wurden, hat die Kombination mit dem aufkommenden Feld der *circuit QED* zur Entstehung eines neuen Forschungsgebiets beigetragen. Die starke Kopplung solcher kohärenter Quantenschaltkreise an Festkörperspinensemble hat neues Licht auf diese Systeme geworfen. In der Tat ist die Physik der kollektiven Kopplung von Resonatoren an inhomogen verbreiterte Spinsysteme noch nicht vollständig verstanden. Bezogen auf diese Arbeit ist insbesondere das überraschende Verhalten des Resonators mit dem geringen Gütefaktor hervorzuheben. Zur Zeit ist die theoretische Forschung damit beschäftigt, den Einfluss inhomogen verbreiterter Spinsysteme auf die Kopplungsphysik zu verstehen. In diesem Zusammenhang wurde bereits gezeigt, dass bestimmte Formen des Spektrums neue Effekte hervorbringen, wie den *cavity protection* Effekt, bei dem die Dephasierung in dunkle Moden unterdrückt wird.

Die Resultate der fokussierten Ionenstrahlimplantation von Erbium Ionen in Y_2SiO_5 Substrate sind aus materialphysikalischer Sicht ebenfalls spannend. Um die inhomogene Linienbreite zu verringern, ist ein detaillierteres Verständnis des Implantationsprozesses und der Art und Weise, wie die Erbiumionen die Yttriumionen ersetzen, notwendig. Zur Verbesserung der Effektivität dieses Prozesses setzen aktuelle Experimente auf Implantation in geheizte Substrate. Sobald fokussierte Ionenstrahlimplantation hochqualitative und stark gekoppelte Spinensembles ermöglicht, wird sie eine Schlüsseltechnologie darstellen auf dem Weg zur praktischen Integration von Spins in *circuit QED* Quantenprozessoren. Eine weitere Möglichkeit bietet sich durch die *on-chip* Integration optischer Wellenleiter auf Erbium implantierte Substrate [SPT13]. Dies könnte kombinierte *on-chip* optische und Mikrowellenspektroskopie ermöglichen und perspektivisch einen *on-chip* Quantenmediakonverter.

Eine spannende Frage ist nun: *Ist es möglich, einen seltenerdbasierten Einzelphotonenspeicher mit Multimodenkapazität zu bauen?* Die Experimente dieser Arbeit zeigen bereits die Multimodenspeicherkapazität des Erbium dotierten Y_2SiO_5. Auch wenn der Aufbau nicht speziell optimiert wurde, erreicht er einen guten Wirkungsgrad. Die Erweiterung der Pulssequenzen auf das Einzelphotonregime kohärenter Pulse ist überschaubar. Das dafür notwendige Signal-zu-Rausch Verhältnis fordert den Einsatz eines kryogenen HEMT Verstärkers mit vorgeschaltetem parametrischem Verstärker. Darüber hinaus ist ein kleinerer Resonator wichtig, da dadurch die für den π-Puls notwendige Maximalleistung herabgesetzt werden kann. Andernfalls würde die hohe Pulsleistung des aktuellen Experiments den HEMT Verstärker zerstören. Um die Effizienz zu steigern, sind einerseits spezielle adiabatische Pulsformen notwendig, die trotz der Magnetfeldanisotropie eine gute Refokussierung ermöglichen [Sig+14]. Andererseits kann die Kohärenz nur durch Reduktion der Ionenkonzentration wesentlich erhöht werden. Innerhalb der nächsten Jahre können diese Verbesserungen durchaus zu Wirkungsgraden von $0,1$ im Wenige- bis Einzelphotonenregime führen [Gre+14]. Parallel dazu ist in näherer Zukunft der Betrieb bei niedrigeren Magnetfeldern basierend auf Er^{3+}:$YAlO_3$ Kristallen und die Integration von Qubits in den Quantenschaltkreis realistisch. Die Gegenwart des Kernspins des Al in $YAlO_3$ könnte möglicherweise eine größere Kohärenzzeit verhindern. Jedoch wäre es möglich, dass durch einen *frozen core* Effekt wie bei Er^{3+}:Y_2SiO_5 dieser Dekohärenzkanal unterdrückt wird. Auf Grundlage der in den letzten Jahren gesammelten Erfahrungen im Bereich der hybriden *circuit QED* Quantensysteme ist ein Prototyp eines Multimoden-Einzelphotonenquantenspeichers sicher innerhalb der nächsten drei bis fünf Jahre realisierbar.

Eine ebenso spannende jedoch schwierige Frage ist: *Ist es möglich, einen Erbium basierten Mikrowellen- zu optischen Photonen-Konverter zu realisieren?* Aus der Kombination von optischer und Mikrowellenspektroskopie bei milli-Kelvin Temperaturen ergeben sich ganz neue Herausforderungen. In dieser Arbeit wurde ein grundlegender Aufbau untersucht, der die Komplexität dieser Aufgabe gut beleuchtet. Die planare Architektur des *on-chip* Resonators führt zu Inhomogenitäten, die spezielle Kontrollpulse erfordern. Ein interessanter Weg ist die Verwendung von 3D Resonatoren, da die größeren geometrischen Abmessungen eine erhöhte Flexibilität bieten. Das erlaubt gleichzeitig einen einfachen optischen Zugang und homogene AC Mikrowellenmagnetfelder. Jedoch verbleibt das Kernproblem der großen inhomogenen Linienbreite des optischen Übergangs. Kürzlich wurde ein theoretisches Konzept zur Einzelphotonen-Konversion in diesem System vorgeschlagen [OBr+14]. Trotz der technischen Herausforderungen wird reversible und kohärente Mikrowellen- zu optischen Photonen-Konvertierung höchstwahrscheinlich möglich werden. Die Frage ist jedoch, ob gute Wirkungsgrade erreicht werden können. Aus der heutigen Sicht sind Architekturen basierend auf nanomechanischen Resonatoren kurzfristig einfacher zu realisieren und im Einzelphotonenregime zu betreiben, da diese ein weniger komplexes Konversionsprotokoll erfordern [And+14].

Das Er:YSO Hybridsystem und vergleichbare, an Mikrowellenschaltkreise gekoppelte Materialien haben ebenfalls Potential in der Forschung an Quantenmetamaterialen. Diese beschäftigt sich mit der Ausbreitung von Licht in künstlichen, gekoppelten Multiresonator- und Multiqubitsystemen. Ein einfaches Metamaterial besteht beispielsweise aus vielen Qubits gekoppelt an einen Mikrowellenleiter [Zag+09]. Dieses kann bereits durch Kopplung eines Er:YSO Kristalls an einen langen planaren Mikrowellenleiter erreicht werden. Kombinierte Multiwellenlängen-Einzelphotonenspektroskopie von Er:YSO oder ähnlichen Materialien kann zur Untersuchung fundamentaler Fragestellungen beitragen, wie beispielsweise kohärentem Quantenenergietransport in einem Multiskalen-Festkörpersystem.

In jedem Falle wird den Hybridsystemen basierend auf Kristallen mit Seltenerddotierung auch weiterhin eine interessante und vielseitige Zukunft bevorstehen.

Bibliography

[Abd+08] Abdufarrukh A. Abdumalikov, Oleg Astafiev, Yasunobu Nakamura, Yuri A. Pashkin, and JawShen Tsai. "Vacuum Rabi splitting due to strong coupling of a flux qubit and a coplanar-waveguide resonator". In: *Physical Review B* 78 (18 Nov. 2008), p. 180502. DOI: `10.1103/PhysRevB.78.180502` (cit. on p. 28).

[Abe+11] Eisuke Abe, Hua Wu, Arzhang Ardavan, and John J. L. Morton. "Electron spin ensemble strongly coupled to a three-dimensional microwave cavity". In: *Applied Physics Letters* 98.25, 251108 (2011), pages. DOI: `10.1063/1.3601930` (cit. on p. 4).

[AB12] A. Abragam and B. Bleaney. *Electron Paramagnetic Resonance of Transition Ions*. Oxford University Press, 2012 (cit. on pp. 7–9, 13, 15, 16, 34).

[Afz+13] M Afzelius, N Sangouard, G Johansson, M U Staudt, and C M Wilson. "Proposal for a coherent quantum memory for propagating microwave photons". In: *New Journal of Physics* 15.6 (2013), p. 065008. DOI: `10.1088/1367-2630/15/6/065008` (cit. on pp. 36, 60, 79).

[Afz+09] Mikael Afzelius, Christoph Simon, Hugues de Riedmatten, and Nicolas Gisin. "Multimode quantum memory based on atomic frequency combs". In: *Physical Review A* 79 (5 May 2009), p. 052329. DOI: `10.1103/PhysRevA.79.052329` (cit. on p. 41).

[Ams+11] R. Amsüss, Ch. Koller, T. Nöbauer, S. Putz, S. Rotter, K. Sandner, S. Schneider, M. Schramböck, G. Steinhauser, H. Ritsch, J. Schmiedmayer, and J. Majer. "Cavity QED with Magnetically Coupled Collective Spin States". In: *Physical Review Letters* 107 (6 Aug. 2011), p. 060502. DOI: `10.1103/PhysRevLett.107.060502` (cit. on p. 3).

[And+14] R. W. Andrews, R. W. Peterson, T. P. Purdy, K. Cicak, R. W. Simmonds, C. A. Regal, and K. W. Lehnert. "Bidirectional and efficient conversion between microwave and optical light". In: *Nature Physics* 10.4 (Apr. 2014), pp. 321–326. DOI: `10.1038/nphys2911` (cit. on pp. 3, 111, 117).

[AR02] G.R. Asatryan and J. Rosa. "EPR of Er3+, Nd3+, and Ce3+ ions in YAlO3 single crystals". English. In: *Physics of the Solid State* 44.5 (2002), pp. 864–869. DOI: `10.1134/1.1477485` (cit. on pp. 64, 66, 68).

[ARM97] H.R. Asatryan, J. Rosa, and J.A. Mareš. "{EPR} studies of Er3+, Nd3+ and Ce3+ in {YAlO3} single crystals". In: *Solid State Communications* 104.1 (1997), pp. 5–9. DOI: `10.1016/S0038-1098(97)00258-5` (cit. on pp. 9, 64).

[BCS57] J. Bardeen, L. N. Cooper, and J. R. Schrieffer. "Microscopic Theory of Superconductivity". In: *Physical Review* 106 (1 Apr. 1957), pp. 162–164. DOI: 10.1103/PhysRev.106.162 (cit. on p. 27).

[Bar+14] R. Barends, J. Kelly, A. Megrant, A. Veitia, D. Sank, E. Jeffrey, T. C. White, J. Mutus, A. G. Fowler, B. Campbell, Y. Chen, Z. Chen, B. Chiaro, A. Dunsworth, C. Neill, P. O/'Malley, P. Roushan, A. Vainsencher, J. Wenner, A. N. Korotkov, A. N. Cleland, and John M. Martinis. "Superconducting quantum circuits at the surface code threshold for fault tolerance". In: *Nature* 508.7497 (Apr. 2014), pp. 500–503. DOI: 10.1038/nature13171 (cit. on p. 27).

[Ben+13] O.W.B. Benningshof, H.R. Mohebbi, I.A.J. Taminiau, G.X. Miao, and D.G. Cory. "Superconducting microstrip resonator for pulsed {ESR} of thin films". In: *Journal of Magnetic Resonance* 230 (2013), pp. 84–87. DOI: 10.1016/j.jmr.2013.01.010 (cit. on p. 97).

[Ber+10] N. Bergeal, F. Schackert, M. Metcalfe, R. Vijay, V. E. Manucharyan, L. Frunzio, D. E. Prober, R. J. Schoelkopf, S. M. Girvin, and M. H. Devoret. "Phase-preserving amplification near the quantum limit with a Josephson ring modulator". In: *Nature* 465.7294 (May 2010), pp. 64–68. DOI: 10.1038/nature09035 (cit. on p. 86).

[Ber+07] BertainaS., GambarelliS., TkachukA., KurkinI. N., MalkinB., StepanovA., and BarbaraB. "Rare-earth solid-state qubits". In: *Nature Nanotechnology* 2.1 (Jan. 2007), pp. 39–42. DOI: 10.1038/nnano.2006.174 (cit. on pp. 4, 68).

[BC08] Pratip Bhattacharyya and Bikas K Chakrabarti. "The mean distance to the n th neighbour in a uniform distribution of random points: an application of probability theory". In: *European Journal of Physics* 29.3 (2008), p. 639. DOI: 10.1088/0143-0807/29/3/023 (cit. on pp. 71, 93).

[Bli+04] B. B. Blinov, D. L. Moehring, L.-M. Duan, and C. Monroe. "Observation of entanglement between a single trapped atom and a single photon". In: *Nature* 428.6979 (Mar. 2004), pp. 153–157. DOI: 10.1038/nature02377 (cit. on p. 2).

[Boe+03] Thomas Boettger, Yongchen Sun, Charles W. Thiel, and Rufus L. Cone. "Material optimization of Er3+:Y2SiO5 at 1.5 μm for optical processing, memory, and laser frequency stabilization applications". In: *Proc. SPIE* 4988 (2003), pp. 51–61. DOI: 10.1117/12.474784 (cit. on p. 24).

[Böt+06a] Thomas Böttger, Y. Sun, C. W. Thiel, and R. L. Cone. "Spectroscopy and dynamics of Er^{3+} : Y_2SiO_5 at 1.5 m". In: *Physical Review B* 74 (7 Aug. 2006), p. 075107. DOI: 10.1103/PhysRevB.74.075107 (cit. on pp. 4, 7).

[Böt+06b] Thomas Böttger, C. W. Thiel, Y. Sun, and R. L. Cone. "Optical decoherence and spectral diffusion at 1.5 m in Er^{3+} : Y_2SiO_5 versus magnetic field, temperature, and Er^{3+} concentration". In: *Physical Review B* 73 (7 Feb. 2006), p. 075101. DOI: `10.1103/PhysRevB.73.075101` (cit. on pp. 7, 9, 22, 24, 25, 68).

[Bou+98] V Bouchiat, D Vion, P Joyez, D Esteve, and M H Devoret. "Quantum coherence with a single Cooper pair". In: *Physica Scripta* 1998.T76 (1998), p. 165. DOI: `10.1238/Physica.Topical.076a00165` (cit. on p. 28).

[Bre09] Claudius Ferdinand Brennecke. "Collective interaction between a Bose-Einstein condensate and a coherent few-photon field". PhD thesis. Eidgenössische Technische Hochschule ETH Zürich, 2009 (cit. on p. 33).

[Bri+09] M G Brik, I Sildos, M Berkowski, and A Suchocki. "Spectroscopic and crystal field studies of $YAlO_3$ single crystals doped with Mn ions". In: *Journal of Physics: Condensed Matter* 21.2 (2009), p. 025404. DOI: `10.1088/0953-8984/21/2/025404` (cit. on p. 9).

[Bro+11] Richard M. Brown, Alexei M. Tyryshkin, Kyriakos Porfyrakis, Erik M. Gauger, Brendon W. Lovett, Arzhang Ardavan, S. A. Lyon, G. Andrew D. Briggs, and John J. L. Morton. "Coherent State Transfer between an Electron and Nuclear Spin in ^{15}**N@C**$_{60}$". In: *Physical Review Letters* 106 (11 Mar. 2011), p. 110504. DOI: `10.1103/PhysRevLett.106.110504` (cit. on pp. 71, 92, 93).

[Bus+11] P. Bushev, A. K. Feofanov, H. Rotzinger, I. Protopopov, J. H. Cole, C. M. Wilson, G. Fischer, A. Lukashenko, and A. V. Ustinov. "Ultralow-power spectroscopy of a rare-earth spin ensemble using a superconducting resonator". In: *Physical Review B* 84 (6 Aug. 2011), p. 060501. DOI: `10.1103/PhysRevB.84.060501` (cit. on pp. 4, 7, 8, 34, 59, 70, 72).

[Bus+14] Félix Bussières, Christoph Clausen, Alexey Tiranov, Boris Korzh, Varun B. Verma, Sae Woo Nam, Francesco Marsili, Alban Ferrier, Philippe Goldner, Harald Herrmann, Christine Silberhorn, Wolfgang Sohler, Mikael Afzelius, and Nicolas Gisin. "Quantum teleportation from a telecom-wavelength photon to a solid-state quantum memory". In: *Nature Photonics* 8 (Sept. 21, 2014), pp. 775–778. DOI: `10.1038/nphoton.2014.215` (cit. on p. 4).

[But30] Stephen Butterworth. "On the Theory of Filter Amplifiers". In: *Wireless Engineer* 7 (1930), pp. 536–541 (cit. on p. 83).

[Che+12] Yu Chen, D. Sank, P. O'Malley, T. White, R. Barends, B. Chiaro, J. Kelly, E. Lucero, M. Mariantoni, A. Megrant, C. Neill, A. Vainsencher, J. Wenner, Y. Yin, A. N. Cleland, and John M. Martinis. "Multiplexed dispersive readout of superconducting phase qubits". In: *Applied Physics Letters* 101.18, 182601 (2012), pages. DOI: `10.1063/1.4764940` (cit. on p. 46).

[Chi+10] I. Chiorescu, N. Groll, S. Bertaina, T. Mori, and S. Miyashita. "Magnetic strong coupling in a spin-photon system and transition to classical regime". In: *Physical Review B* 82 (2 July 2010), p. 024413. DOI: 10.1103/PhysRevB.82.024413 (cit. on p. 4).

[Cir+97] J. I. Cirac, P. Zoller, H. J. Kimble, and H. Mabuchi. "Quantum State Transfer and Entanglement Distribution among Distant Nodes in a Quantum Network". In: *Physical Review Letters* 78 (16 Apr. 1997), pp. 3221–3224. DOI: 10.1103/PhysRevLett.78.3221 (cit. on p. 2).

[CW08] John Clarke and Frank K. Wilhelm. "Superconducting quantum bits". In: *Nature* 453.7198 (June 2008), pp. 1031–1042. DOI: 10.1038/nature07128 (cit. on p. 3).

[Cla+11] Christoph Clausen, Imam Usmani, Felix Bussieres, Nicolas Sangouard, Mikael Afzelius, Hugues de Riedmatten, and Nicolas Gisin. "Quantum storage of photonic entanglement in a crystal". In: *Nature* 469.7331 (Jan. 2011), pp. 508–511. DOI: 10.1038/nature09662 (cit. on p. 3).

[Cla+13] Conrad Clauss, Daniel Bothner, Dieter Koelle, Reinhold Kleiner, Lapo Bogani, Marc Scheffler, and Martin Dressel. "Broadband electron spin resonance from 500 MHz to 40 GHz using superconducting coplanar waveguides". In: *Applied Physics Letters* 102.16, 162601 (2013), pages. DOI: 10.1063/1.4802956 (cit. on pp. 38, 60).

[De +12] Kristiaan De Greve, Leo Yu, Peter L. McMahon, Jason S. Pelc, Chandra M. Natarajan, Na Young Kim, Eisuke Abe, Sebastian Maier, Christian Schneider, Martin Kamp, Sven Hofling, Robert H. Hadfield, Alfred Forchel, M. M. Fejer, and Yoshihisa Yamamoto. "Quantum-dot spin-photon entanglement via frequency downconversion to telecom wavelength". In: *Nature* 491.7424 (Nov. 2012), pp. 421–425. DOI: 10.1038/nature11577 (cit. on p. 3).

[DS13] M. H. Devoret and R. J. Schoelkopf. "Superconducting Circuits for Quantum Information: An Outlook". In: *Science* 339.6124 (2013), pp. 1169–1174. DOI: 10.1126/science.1231930 (cit. on p. 27).

[Din+11] I. Diniz, S. Portolan, R. Ferreira, J. M. Gérard, P. Bertet, and A. Auffèves. "Strongly coupling a cavity to inhomogeneous ensembles of emitters: Potential for long-lived solid-state quantum memories". In: *Physical Review A* 84 (6 Dec. 2011), p. 063810. DOI: 10.1103/PhysRevA.84.063810 (cit. on pp. 35, 36).

[DiV00] David P. DiVincenzo. "The Physical Implementation of Quantum Computation". In: *Fortschritte der Physik* 48.9-11 (2000), pp. 771–783 (cit. on p. 2).

[Dol+11] F. Dolde, H. Fedder, M. W. Doherty, T. Nobauer, F. Rempp, G. Balasubramanian, T. Wolf, F. Reinhard, L. C. L. Hollenberg, F. Jelezko, and J. Wrachtrup. "Electric-field sensing using single diamond spins". In: *Nature Physics* 7.6 (June 2011), pp. 459–463. DOI: 10.1038/nphys1969 (cit. on p. 97).

[Dou+95] J L Doualan, C Labbe, P Le Boulanger, J Margerie, R Moncorge, and H Timonen. "Energy levels of the laser active Er 3+ ion in each of the two crystallographic sites of yttrium orthosilicate". In: *Journal of Physics: Condensed Matter* 7.26 (1995), p. 5111. DOI: 10.1088/0953-8984/7/26/017 (cit. on pp. 8, 16, 106).

[DS88] R. Dupree and M.E. Smith. "Structural influences on high-resolution yttrium-89 {NMR} spectra of solids". In: *Chemical Physics Letters* 148.1 (1988), pp. 41–44. DOI: 10.1016/0009-2614(88)87257-9 (cit. on p. 89).

[Fin+08] J. M. Fink, M. Goppl, M. Baur, R. Bianchetti, P. J. Leek, A. Blais, and A. Wallraff. "Climbing the Jaynes-Cummings ladder and observing its nonlinearity in a cavity QED system". In: *Nature* 454.7202 (July 2008), pp. 315–318. DOI: 10.1038/nature07112 (cit. on p. 33).

[Fri+00] Jonathan R. Friedman, Vijay Patel, W. Chen, S. K. Tolpygo, and J. E. Lukens. "Quantum superposition of distinct macroscopic states". In: *Nature* 406.6791 (July 2000), pp. 43–46. DOI: 10.1038/35017505 (cit. on p. 28).

[Gao08] Jiansong Gao. "The physics of superconducting microwave resonators". PhD thesis. California Institute of Technology, 2008 (cit. on p. 31).

[Gao+08a] Jiansong Gao, Miguel Daal, John M. Martinis, Anastasios Vayonakis, Jonas Zmuidzinas, Bernard Sadoulet, Benjamin A. Mazin, Peter K. Day, and Henry G. Leduc. "A semiempirical model for two-level system noise in superconducting microresonators". In: *Applied Physics Letters* 92.21 (2008), p. 212504. DOI: 10.1063/1.2937855 (cit. on p. 46).

[Gao+08b] Jiansong Gao, Miguel Daal, Anastasios Vayonakis, Shwetank Kumar, Jonas Zmuidzinas, Bernard Sadoulet, Benjamin A. Mazin, Peter K. Day, and Henry G. Leduc. "Experimental evidence for a surface distribution of two-level systems in superconducting lithographed microwave resonators". In: *Applied Physics Letters* 92.15 (2008), p. 152505. DOI: 10.1063/1.2906373 (cit. on p. 46).

[Gao+07] Jiansong Gao, Jonas Zmuidzinas, Benjamin A. Mazin, Henry G. LeDuc, and Peter K. Day. "Noise properties of superconducting coplanar waveguide microwave resonators". In: *Applied Physics Letters* 90.10 (2007), p. 102507. DOI: 10.1063/1.2711770 (cit. on p. 46).

[GC08] J. C. Garrison and R. Y. Chiao. *Quantum Optics*. Oxford University Press, 2008 (cit. on pp. 23, 35).

[Geo+10] Richard E. George, Wayne Witzel, H. Riemann, N. V. Abrosimov, N. Nötzel, Mike L. W. Thewalt, and John J. L. Morton. "Electron Spin Coherence and Electron Nuclear Double Resonance of Bi Donors in Natural Si". In: *Physical Review Letters* 105 (6 Aug. 2010), p. 067601. DOI: 10.1103/PhysRevLett.105.067601 (cit. on p. 4).

[GAN14] I. M. Georgescu, S. Ashhab, and Franco Nori. "Quantum simulation". In: *Rev. Mod. Phys.* 86 (1 Mar. 2014), pp. 153–185. DOI: 10.1103/RevModPhys.86.153 (cit. on p. 27).

[GL50] V. L. Ginzburg and L. D. Landau. "On the theory of superconductivity". In: *Zh. Eksperim. i. Teor. Fiz.* 20 (1950), p. 1064 (cit. on p. 27).

[GT07] Nicolas Gisin and Rob Thew. "Quantum communication". In: *Nat Photon* 1.3 (Mar. 2007), pp. 165–171. DOI: 10.1038/nphoton.2007.22 (cit. on p. 2).

[Gor+14a] Maxim Goryachev, Warrick G. Farr, Natalia C. Carvalho, Daniel L. Creedon, Jean-Michel Le Floch, Sebastian Probst, Pavel Bushev, and Michael E. Tobar. *Strong Coupling Between Whispering Gallery Photons and Spin States of Iron Group Impurity Ions*. 2014 (cit. on p. 4).

[Gor+14b] Maxim Goryachev, Warrick G. Farr, Daniel L. Creedon, Yaohui Fan, Mikhail Kostylev, and Michael E. Tobar. "High-Cooperativity Cavity QED with Magnons at Microwave Frequencies". In: *Physical Review Applied* 2 (5 Nov. 2014), p. 054002. DOI: 10.1103/PhysRevApplied.2.054002 (cit. on p. 4).

[Gre+14] C. Grezes, B. Julsgaard, Y. Kubo, M. Stern, T. Umeda, J. Isoya, H. Sumiya, H. Abe, S. Onoda, T. Ohshima, V. Jacques, J. Esteve, D. Vion, D. Esteve, K. Mølmer, and P. Bertet. "Multimode Storage and Retrieval of Microwave Fields in a Spin Ensemble". In: *Physical Review X* 4 (2 June 2014), p. 021049. DOI: 10.1103/PhysRevX.4.021049 (cit. on pp. 3, 94, 97, 109, 110, 115, 117).

[Gui+06] O. Guillot-Noël, Ph. Goldner, Y. Le Du, E. Baldit, P. Monnier, and K. Bencheikh. "Hyperfine interaction of Er^{3+} ions in Y_2SiO_5: An electron paramagnetic resonance spectroscopy study". In: *Physical Review B* 74 (21 Dec. 2006), p. 214409. DOI: 10.1103/PhysRevB.74.214409 (cit. on pp. 8, 9, 12, 59, 62).

[Gui+07] O. Guillot-Noël, H. Vezin, Ph. Goldner, F. Beaudoux, J. Vincent, J. Lejay, and I. Lorgeré. "Direct observation of rare-earth-host interactions in Er : Y_2SiO_5". In: *Physical Review B* 76 (18 Nov. 2007), p. 180408. DOI: 10.1103/PhysRevB.76.180408 (cit. on pp. 22, 88).

[HGA06] R. Hanson, O. Gywat, and D. D. Awschalom. "Room-temperature manipulation and decoherence of a single spin in diamond". In: *Physical Review B* 74 (16 Oct. 2006), p. 161203. DOI: 10.1103/PhysRevB.74.161203 (cit. on p. 97).

[HR13] Serge Haroche and Jean-Michel Raimond. *Exploring the Quantum*. Ed. by Oxford University Press. 2013 (cit. on p. 31).

[HL11] W. M. Haynes and David R. Lide, eds. *CRC Handbook of Chemistry and Physics*. CRC Press, 2011 (cit. on pp. 66, 67).

[Hét+08] G. Hétet, J. J. Longdell, A. L. Alexander, P. K. Lam, and M. J. Sellars. "Electro-Optic Quantum Memory for Light Using Two-Level Atoms". In: *Physical Review Letters* 100 (2 Jan. 2008), p. 023601. DOI: 10.1103/PhysRevLett.100.023601 (cit. on p. 40).

[Hof89] A. J. Hoff. *Advanced EPR : applications in biology and biochemistry*. Amsterdam; New York; New York, NY, U.S.A.: Elsevier ; Distributors for the U.S. and Canada, Elsevier Science Pub. Co., 1989, pages (cit. on p. 21).

[Hof+09] Max Hofheinz, H. Wang, M. Ansmann, Radoslaw C. Bialczak, Erik Lucero, M. Neeley, A. D. O'Connell, D. Sank, J. Wenner, John M. Martinis, and A. N. Cleland. "Synthesizing arbitrary quantum states in a superconducting resonator". In: *Nature* 459.7246 (May 2009), pp. 546–549. DOI: `10.1038/nature08005` (cit. on pp. 3, 27).

[Hof+12] Julian Hofmann, Michael Krug, Norbert Ortegel, Lea Gérard, Markus Weber, Wenjamin Rosenfeld, and Harald Weinfurter. "Heralded Entanglement Between Widely Separated Atoms". In: *Science* 337.6090 (2012), pp. 72–75. DOI: `10.1126/science.1221856` (cit. on p. 3).

[HP40] T. Holstein and H. Primakoff. "Field Dependence of the Intrinsic Domain Magnetization of a Ferromagnet". In: *Physical Review* 58 (12 Dec. 1940), pp. 1098–1113. DOI: `10.1103/PhysRev.58.1098` (cit. on p. 35).

[Hul05] Robert Hull. *Spectroscopic Properties of Rare Earths in Optical Materials*. Ed. by Jürgen Parisi, R. M. Jr Osgood, Hans Warlimont, Guokui Liu, and Bernard Jacquier. Springer Series in Materials Science. Berlin, Heidelberg: Springer Berlin Heidelberg, 2005 (cit. on p. 89).

[Ima09] Atac Imamoğlu. "Cavity QED Based on Collective Magnetic Dipole Coupling: Spin Ensembles as Hybrid Two-Level Systems". In: *Physical Review Letters* 102 (8 Feb. 2009), p. 083602. DOI: `10.1103/PhysRevLett.102.083602` (cit. on p. 3).

[Jer+12] M. Jerger, S. Poletto, P. Macha, U. Hübner, E. Il'ichev, and A. V. Ustinov. "Frequency division multiplexing readout and simultaneous manipulation of an array of flux qubits". In: *Applied Physics Letters* 101.4 (2012), p. 042604. DOI: `10.1063/1.4739454` (cit. on p. 46).

[Jer13] Markus Jerger. "Experiments on Superconducting Qubits Coupled to Resonators". PhD thesis. Karlsruhe Institute of Technology (KIT), 2013 (cit. on p. 29).

[Jos62] B.D. Josephson. "Possible new effects in superconductive tunnelling". In: *Physics Letters* 1.7 (1962), pp. 251–253. DOI: `10.1016/0031-9163(62)91369-0` (cit. on p. 28).

[Kha+12] M. S. Khalil, M. J. A. Stoutimore, F. C. Wellstood, and K. D. Osborn. "An analysis method for asymmetric resonator transmission applied to superconducting devices". In: *Journal of Applied Physics* 111.5 (2012), pp. 054510+. DOI: `10.1063/1.3692073` (cit. on p. 31).

[Kim08] H. J. Kimble. "The quantum internet". In: *Nature* 453.7198 (June 2008), pp. 1023–1030. DOI: `10.1038/nature07127` (cit. on p. 2).

[KA62] J. R. Klauder and P. W. Anderson. "Spectral Diffusion Decay in Spin Resonance Experiments". In: *Physical Review* 125 (3 Feb. 1962), pp. 912–932. DOI: `10.1103/PhysRev.125.912` (cit. on pp. 91, 92).

[Koc+07] Jens Koch, Terri M. Yu, Jay Gambetta, A. A. Houck, D. I. Schuster, J. Majer, Alexandre Blais, M. H. Devoret, S. M. Girvin, and R. J. Schoelkopf. "Charge-insensitive qubit design derived from the Cooper pair box". In: *Physical Review A* 76 (4 Oct. 2007), p. 042319. DOI: 10.1103/PhysRevA.76.042319 (cit. on p. 28).

[Kol+12] R. Kolesov, K. Xia, R. Reuter, R. Stöhr, A. Zappe, J. Meijer, P.R. Hemmer, and J. Wrachtrup. "Optical detection of a single rare-earth ion in a crystal". In: *Nat Commun* 3 (Aug. 2012), pp. 1029–. DOI: 10.1038/ncomms2034 (cit. on p. 68).

[Kri+14] Dmitry O. Krimer, Stefan Putz, Johannes Majer, and Stefan Rotter. "Non-Markovian dynamics of a single-mode cavity strongly coupled to an inhomogeneously broadened spin ensemble". In: *Physical Review A* 90 (4 Oct. 2014), p. 043852. DOI: 10.1103/PhysRevA.90.043852 (cit. on pp. 37, 82, 85).

[Kub+12] Y. Kubo, I. Diniz, C. Grezes, T. Umeda, J. Isoya, H. Sumiya, T. Yamamoto, H. Abe, S. Onoda, T. Ohshima, V. Jacques, A. Dréau, J.-F. Roch, A. Auffeves, D. Vion, D. Esteve, and P. Bertet. "Electron spin resonance detected by a superconducting qubit". In: *Physical Review B* 86 (6 Aug. 2012), p. 064514. DOI: 10.1103/PhysRevB.86.064514 (cit. on pp. 3, 70).

[Kub+11] Y. Kubo, C. Grezes, A. Dewes, T. Umeda, J. Isoya, H. Sumiya, N. Morishita, H. Abe, S. Onoda, T. Ohshima, V. Jacques, A. Dréau, J.-F. Roch, I. Diniz, A. Auffeves, D. Vion, D. Esteve, and P. Bertet. "Hybrid Quantum Circuit with a Superconducting Qubit Coupled to a Spin Ensemble". In: *Physical Review Letters* 107 (22 Nov. 2011), p. 220501. DOI: 10.1103/PhysRevLett.107.220501 (cit. on pp. 3, 41).

[Kub+10] Y. Kubo, F. R. Ong, P. Bertet, D. Vion, V. Jacques, D. Zheng, A. Dréau, J.-F. Roch, A. Auffeves, F. Jelezko, J. Wrachtrup, M. F. Barthe, P. Bergonzo, and D. Esteve. "Strong Coupling of a Spin Ensemble to a Superconducting Resonator". In: *Physical Review Letters* 105 (14 Sept. 2010), p. 140502. DOI: 10.1103/PhysRevLett.105.140502 (cit. on p. 3).

[Kuk14] Nadezhda Kukharchyk. (unpublished). PhD thesis. Ruhr Universität Bochum, 2014 (cit. on p. 69).

[Kuk+14] Nadezhda Kukharchyk, Shovon Pal, Jasper Rödiger, Arne Ludwig, Sebastian Probst, Alexey V. Ustinov, Pavel Bushev, and Andreas D. Wieck. "Photoluminescence of focused ion beam implanted Er3+:Y2SiO5 crystals". In: *physica status solidi (RRL) - Rapid Research Letters* 8.10 (2014), pp. 880–884. DOI: 10.1002/pssr.201409304 (cit. on pp. 68, 69, 107, 114).

[KC80] I.N. Kurkin and K.P. Chernov. "{EPR} and spin-lattice relaxation of rare-earth activated centres in {Y2SiO5} single crystals". In: *Physica B+C* 101.2 (1980), pp. 233–238. DOI: 10.1016/0378-4363(80)90107-2 (cit. on pp. 4, 9, 61).

[Kut+95] C. Kutter, H. P. Moll, J. van Tol, H. Zuckermann, J. C. Maan, and P. Wyder. "Electron-Spin Echoes at 604 GHz Using Far Infrared Lasers". In: *Physical Review Letters* 74 (15 Apr. 1995), pp. 2925–2928. DOI: `10.1103/PhysRevLett.74.2925` (cit. on p. 91).

[LWM92] C. Li, C. Wyon, and R. Moncorge. "Spectroscopic properties and fluorescence dynamics of Er3+ and Yb3+ in Y2SiO5". In: *Quantum Electronics, IEEE Journal of* 28.4 (Apr. 1992), pp. 1209–1221. DOI: `10.1109/3.135248` (cit. on p. 8).

[Luc+12] Erik Lucero, R. Barends, Y. Chen, J. Kelly, M. Mariantoni, A. Megrant, P. O/'Malley, D. Sank, A. Vainsencher, J. Wenner, T. White, Y. Yin, A. N. Cleland, and John M. Martinis. "Computing prime factors with a Josephson phase qubit quantum processor". In: *Nature Physics* 8.10 (Oct. 2012), pp. 719–723. DOI: `10.1038/nphys2385` (cit. on pp. 3, 27).

[Mac+97] R. M. Macfarlane, T. L. Harris, Y. Sun, R. L. Cone, and R. W. Equall. "Measurement of photon echoes in ErY2SiO5 at 1.5 μm with a diode laser and an amplifier". In: *Opt. Lett.* 22.12 (June 1997), pp. 871–873. DOI: `10.1364/OL.22.000871` (cit. on p. 24).

[Man+09] Vladimir E. Manucharyan, Jens Koch, Leonid I. Glazman, and Michel H. Devoret. "Fluxonium: Single Cooper-Pair Circuit Free of Charge Offsets". In: *Science* 326.5949 (2009), pp. 113–116. DOI: `10.1126/science.1175552` (cit. on p. 28).

[Mar+10] D. Marcos, M. Wubs, J. M. Taylor, R. Aguado, M. D. Lukin, and A. S. Sørensen. "Coupling Nitrogen-Vacancy Centers in Diamond to Superconducting Flux Qubits". In: *Physical Review Letters* 105 (21 Nov. 2010), p. 210501. DOI: `10.1103/PhysRevLett.105.210501` (cit. on p. 64).

[MDC85] John M. Martinis, Michel H. Devoret, and John Clarke. "Energy-Level Quantization in the Zero-Voltage State of a Current-Biased Josephson Junction". In: *Physical Review Letters* 55 (15 Oct. 1985), pp. 1543–1546. DOI: `10.1103/PhysRevLett.55.1543` (cit. on p. 28).

[Mau+12] P. C. Maurer, G. Kucsko, C. Latta, L. Jiang, N. Y. Yao, S. D. Bennett, F. Pastawski, D. Hunger, N. Chisholm, M. Markham, D. J. Twitchen, J. I. Cirac, and M. D. Lukin. "Room-Temperature Quantum Bit Memory Exceeding One Second". In: *Science* 336.6086 (2012), pp. 1283–1286. DOI: `10.1126/science.1220513` (cit. on p. 2).

[McA+12] D. L. McAuslan, J. G. Bartholomew, M. J. Sellars, and J. J. Longdell. "Reducing decoherence in optical and spin transitions in rare-earth-metal-ion–doped materials". In: *Physical Review A* 85 (3 Mar. 2012), p. 032339. DOI: `10.1103/PhysRevA.85.032339` (cit. on pp. 3, 64).

[MO33] W. Meissner and R. Ochsenfeld. "Ein neuer Effekt bei Eintritt der Supraleitfähigkeit". In: *Naturwissenschaften* 21 (1933), p. 787 (cit. on p. 27).

[Mel+02] A. Melnikov, T. Gerya, M. Hillmann, I. Kamphausen, W. Oswald, P. Stauche, R. Wernhardt, and A.D. Wieck. "Development of an Au-Dy-Si liquid alloy ion source for focussed ion beam implantation". In: *Nuclear Instruments and Methods in Physics Research Section B: Beam Interactions with Materials and Atoms* 195.3-4 (2002), pp. 422–425. DOI: 10.1016/S0168-583X(02)01127-8 (cit. on p. 69).

[MBB67] C Michel, G Buisson, and E F Bertaut. "Structure de Y2 Si O5". In: *Comptes Rendus Hebdomadaires des Seances de l'Academie des Sciences, Serie B* 264 (1967), pp. 397–399 (cit. on p. 89).

[Mim68] W. B. Mims. "Phase Memory in Electron Spin Echoes, Lattice Relaxation Effects in $CaWO_4$: Er, Ce, Mn". In: *Physical Review* 168 (2 Apr. 1968), pp. 370–389. DOI: 10.1103/PhysRev.168.370 (cit. on p. 22).

[Moo+99] J. E. Mooij, T. P. Orlando, L. Levitov, Lin Tian, Caspar H. van der Wal, and Seth Lloyd. "Josephson Persistent-Current Qubit". In: *Science* 285.5430 (1999), pp. 1036–1039. DOI: 10.1126/science.285.5430.1036 (cit. on pp. 28, 29).

[NPT99] Y. Nakamura, Yu. A. Pashkin, and J. S. Tsai. "Coherent control of macroscopic quantum states in a single-Cooper-pair box". In: *Nature* 398.6730 (Apr. 1999), pp. 786–788. DOI: 10.1038/19718 (cit. on p. 28).

[Nee+08] Matthew Neeley, M. Ansmann, Radoslaw C. Bialczak, M. Hofheinz, N. Katz, Erik Lucero, A. O/'Connell, H. Wang, A. N. Cleland, and John M. Martinis. "Process tomography of quantum memory in a Josephson-phase qubit coupled to a two-level state". In: *Nature Physics* 4.7 (July 2008), pp. 523–526. DOI: 10.1038/nphys972 (cit. on p. 2).

[OBr+14] Christopher O'Brien, Nikolai Lauk, Susanne Blum, Giovanna Morigi, and Michael Fleischhauer. "Interfacing Superconducting Qubits and Telecom Photons via a Rare-Earth-Doped Crystal". In: *Physical Review Letters* 113 (6 Aug. 2014), p. 063603. DOI: 10.1103/PhysRevLett.113.063603 (cit. on pp. 3, 4, 40, 73, 111, 117).

[OCo+10] A. D. O'Connell, M. Hofheinz, M. Ansmann, Radoslaw C. Bialczak, M. Lenander, Erik Lucero, M. Neeley, D. Sank, H. Wang, M. Weides, J. Wenner, John M. Martinis, and A. N. Cleland. "Quantum ground state and single-phonon control of a mechanical resonator". In: *Nature* 464.7289 (Apr. 2010), pp. 697–703. DOI: 10.1038/nature08967 (cit. on p. 3).

[Onn11] H. K. Onnes. "The resistance of pure mercury at helium temperatures". In: *Commun. Phys. Lab. Univ. Leiden* 12 (1911), p. 120 (cit. on p. 27).

[Pai+11] Hanhee Paik, D. I. Schuster, Lev S. Bishop, G. Kirchmair, G. Catelani, A. P. Sears, B. R. Johnson, M. J. Reagor, L. Frunzio, L. I. Glazman, S. M. Girvin, M. H. Devoret, and R. J. Schoelkopf. "Observation of High Coherence in Josephson Junction Qubits Measured in a Three-Dimensional Circuit QED Architecture". In: *Physical Review Letters* 107 (24 Dec. 2011), p. 240501. DOI: 10.1103/PhysRevLett.107.240501 (cit. on p. 3).

[Pet+09] David Petrosyan, Guy Bensky, Gershon Kurizki, Igor Mazets, Johannes Majer, and Jörg Schmiedmayer. "Reversible state transfer between superconducting qubits and atomic ensembles". In: *Physical Review A* 79 (4 Apr. 2009), p. 040304. DOI: 10.1103/PhysRevA.79.040304 (cit. on p. 3).

[Poz05] David M. Pozar. *Microwave Engineering*. John Wiley & Sons, Inc., 2005 (cit. on p. 30).

[Pro+14a] S. Probst, N. Kukharchyk, H. Rotzinger, A. Tkalčec, S. Wünsch, A. D. Wieck, M. Siegel, A. V. Ustinov, and P. A. Bushev. "Hybrid quantum circuit with implanted erbium ions". In: *Applied Physics Letters* 105.16, 162404 (2014), pages. DOI: 10.1063/1.4898696 (cit. on pp. 68, 97, 108, 114).

[Pro+13] S. Probst, H. Rotzinger, S. Wünsch, P. Jung, M. Jerger, M. Siegel, A. V. Ustinov, and P. A. Bushev. "Anisotropic Rare-Earth Spin Ensemble Strongly Coupled to a Superconducting Resonator". In: *Physical Review Letters* 110 (15 Apr. 2013), p. 157001. DOI: 10.1103/PhysRevLett.110.157001 (cit. on pp. 4, 38, 56, 70, 73, 107, 113).

[Pro+14b] S. Probst, F. B. Song, P. A. Bushev, A. V. Ustinov, and M. Weides. *Efficient and robust analysis of complex scattering data under noise in microwave resonators*. 2014 (cit. on pp. 31, 46, 52).

[Pro+14c] S. Probst, A. Tkalčec, H. Rotzinger, D. Rieger, J-M. Le Floch, M. Goryachev, M. E. Tobar, A. V. Ustinov, and P. A. Bushev. "Three-dimensional cavity quantum electrodynamics with a rare-earth spin ensemble". In: *Physical Review B* 90 (10 Sept. 2014), 100404(R). DOI: 10.1103/PhysRevB.90.100404 (cit. on pp. 4, 75, 76, 108, 114).

[Put+14] S. Putz, D. O. Krimer, R. Amsuss, A. Valookaran, T. Nobauer, J. Schmiedmayer, S. Rotter, and J. Majer. "Protecting a spin ensemble against decoherence in the strong-coupling regime of cavity QED". In: *Nature Physics* advance online publication (Aug. 2014), pages (cit. on pp. 3, 14, 37, 82, 85).

[Rab+06] P. Rabl, D. DeMille, J. M. Doyle, M. D. Lukin, R. J. Schoelkopf, and P. Zoller. "Hybrid Quantum Processors: Molecular Ensembles as Quantum Memory for Solid State Circuits". In: *Physical Review Letters* 97 (3 July 2006), p. 033003. DOI: 10.1103/PhysRevLett.97.033003 (cit. on p. 3).

[Rak+09] R. M. Rakhmatullin, I. N. Kurkin, G. V. Mamin, S. B. Orlinskii, M. R. Gafurov, E. I. Baibekov, B. Z. Malkin, S. Gambarelli, S. Bertaina, and B. Barbara. "Coherent spin manipulations in Yb^{3+} : $CaWO_4$ at X- and W-band EPR frequencies". In: *Physical Review B* 79 (17 May 2009), p. 172408. DOI: 10.1103/PhysRevB.79.172408 (cit. on pp. 4, 68).

[Ree+12] M. D. Reed, L. DiCarlo, S. E. Nigg, L. Sun, L. Frunzio, S. M. Girvin, and R. J. Schoelkopf. "Realization of three-qubit quantum error correction with superconducting circuits". In: *Nature* 482.7385 (Feb. 2012), pp. 382–385 (cit. on p. 3).

[Rie13] D. Rieger. "Optical Spectroscopy of an Erbium-Doped Crystal". MA thesis. Physikalisches Institut, Karlsruhe Institute of Technology (KIT), 2013 (cit. on pp. 24, 101, 104).

[Rig+12] Chad Rigetti, Jay M. Gambetta, Stefano Poletto, B. L. T. Plourde, Jerry M. Chow, A. D. Córcoles, John A. Smolin, Seth T. Merkel, J. R. Rozen, George A. Keefe, Mary B. Rothwell, Mark B. Ketchen, and M. Steffen. "Superconducting qubit in a waveguide cavity with a coherence time approaching 0.1 ms". In: *Physical Review B* 86 (10 Sept. 2012), p. 100506. DOI: 10.1103/PhysRevB.86.100506 (cit. on p. 3).

[Rit+12] Stephan Ritter, Christian Nolleke, Carolin Hahn, Andreas Reiserer, Andreas Neuzner, Manuel Uphoff, Martin Mucke, Eden Figueroa, Joerg Bochmann, and Gerhard Rempe. "An elementary quantum network of single atoms in optical cavities". In: *Nature* 484.7393 (Apr. 2012), pp. 195–200. DOI: 10.1038/nature11023 (cit. on p. 3).

[Ros+07] Wenjamin Rosenfeld, Stefan Berner, Jürgen Volz, Markus Weber, and Harald Weinfurter. "Remote Preparation of an Atomic Quantum Memory". In: *Physical Review Letters* 98 (5 Feb. 2007), p. 050504. DOI: 10.1103/PhysRevLett.98.050504 (cit. on p. 2).

[Sag+11] Erhan Saglamyurek, Neil Sinclair, Jeongwan Jin, Joshua A. Slater, Daniel Oblak, Felix Bussieres, Mathew George, Raimund Ricken, Wolfgang Sohler, and Wolfgang Tittel. "Broadband waveguide quantum memory for entangled photons". In: *Nature* 469.7331 (Jan. 2011), pp. 512–515. DOI: 10.1038/nature09719 (cit. on p. 3).

[San+12] K. Sandner, H. Ritsch, R. Amsüss, Ch. Koller, T. Nöbauer, S. Putz, J. Schmiedmayer, and J. Majer. "Strong magnetic coupling of an inhomogeneous nitrogen-vacancy ensemble to a cavity". In: *Physical Review A* 85 (5 May 2012), p. 053806. DOI: 10.1103/PhysRevA.85.053806 (cit. on pp. 3, 14).

[SPT13] C. Schuck, W. H. P. Pernice, and H. X. Tang. "NbTiN superconducting nanowire detectors for visible and telecom wavelengths single photon counting on Si3N4 photonic circuits". In: *Applied Physics Letters* 102.5, 051101 (2013), pages. DOI: 10.1063/1.4788931 (cit. on pp. 73, 110, 116).

[Sch+07] D. I. Schuster, A. A. Houck, J. A. Schreier, A. Wallraff, J. M. Gambetta, A. Blais, L. Frunzio, J. Majer, B. Johnson, M. H. Devoret, S. M. Girvin, and R. J. Schoelkopf. "Resolving photon number states in a superconducting circuit". In: *Nature* 445.7127 (Feb. 2007), pp. 515–518. DOI: 10.1038/nature05461 (cit. on p. 28).

[Sch+10] D. I. Schuster, A. P. Sears, E. Ginossar, L. DiCarlo, L. Frunzio, J. J. L. Morton, H. Wu, G. A. D. Briggs, B. B. Buckley, D. D. Awschalom, and R. J. Schoelkopf. "High-Cooperativity Coupling of Electron-Spin Ensembles to Superconducting Cavities". In: *Physical Review Letters* 105 (14 Sept. 2010), p. 140501. DOI: 10.1103/PhysRevLett.105.140501 (cit. on pp. 3, 37, 60, 70).

[SE01] A. Schweiger and G. Eschke. *Principles of pulse elecrton paramagnetic resonance*. Oxford, New York: Oxford University Press, 2001 (cit. on pp. 9, 15, 19, 22, 38, 69, 92).

[Sco07] Lukehart Charles M. Scott Robert A. *Applications of physical methods to inorganic and bioinorganic chemistry*. Hoboken, NJ: Wiley, 2007, pages (cit. on p. 21).

[Sen+08] S. Sendelbach, D. Hover, A. Kittel, M. Mück, John M. Martinis, and R. McDermott. "Magnetism in SQUIDs at Millikelvin Temperatures". In: *Physical Review Letters* 100 (22 June 2008), p. 227006. DOI: 10.1103/PhysRevLett.100.227006 (cit. on p. 64).

[Sig+14] A. J. Sigillito, H. Malissa, A. M. Tyryshkin, H. Riemann, N. V. Abrosimov, P. Becker, H.-J. Pohl, M. L. W. Thewalt, K. M. Itoh, J. J. L. Morton, A. A. Houck, D. I. Schuster, and S. A. Lyon. "Fast, low-power manipulation of spin ensembles in superconducting microresonators". In: *Applied Physics Letters* 104.22, 222407 (2014), pages. DOI: 10.1063/1.4881613 (cit. on pp. 70, 75, 94, 97, 104, 110, 117).

[Sim+10] Simon, C., Afzelius, M., Appel, J., Boyer de la Giroday, A., Dewhurst, S. J., Gisin, N., Hu, C. Y., Jelezko, F., Kröll, S., Müller, J. H., Nunn, J., Polzik, E. S., Rarity, J. G., De Riedmatten, H., Rosenfeld, W., Shields, A. J., Sköld, N., Stevenson, R. M., Thew, R., Walmsley, I. A., Weber, M. C., Weinfurter, H., Wrachtrup, J., and Young, R. J. "Quantum memories". In: *Eur. Phys. J. D* 58.1 (2010), pp. 1–22. DOI: 10.1140/epjd/e2010-00103-y (cit. on p. 2).

[Spa+13] B M Sparkes, J Bernu, M Hosseini, J Geng, Q Glorieux, P A Altin, P K Lam, N P Robins, and B C Buchler. "Gradient echo memory in an ultra-high optical depth cold atomic ensemble". In: *New Journal of Physics* 15.8 (2013), p. 085027. DOI: 10.1088/1367-2630/15/8/085027 (cit. on p. 40).

[Sta+12] Matthias U Staudt, Io-Chun Hoi, Philip Krantz, Martin Sandberg, Michaël Simoen, Pavel Bushev, Nicolas Sangouard, Mikael Afzelius, Vitaly S Shumeiko, Göran Johansson, Per Delsing, and C M Wilson. "Coupling of an erbium spin ensemble to a superconducting resonator". In: *Journal of Physics B: Atomic, Molecular and Optical Physics* 45.12 (2012), p. 124019. DOI: 10.1088/0953-4075/45/12/124019 (cit. on p. 4).

[Ste+12] M. Steger, K. Saeedi, M. L. W. Thewalt, J. J. L. Morton, H. Riemann, N. V. Abrosimov, P. Becker, and H.-J. Pohl. "Quantum Information Storage for over 180 s Using Donor Spins in a 28Si 'Semiconductor Vacuum'". In: *Science* 336.6086 (2012), pp. 1280–1283. DOI: 10.1126/science.1217635 (cit. on p. 2).

[Ste+14] M. Stern, G. Catelani, Y. Kubo, C. Grezes, A. Bienfait, D. Vion, D. Esteve, and P. Bertet. "Flux Qubits with Long Coherence Times for Hybrid Quantum Circuits". In: *Physical Review Letters* 113 (12 Sept. 2014), p. 123601. DOI: 10.1103/PhysRevLett.113.123601 (cit. on p. 32).

[SS06] Stefan Stoll and Arthur Schweiger. "EasySpin, a comprehensive software package for spectral simulation and analysis in {EPR}". In: *Journal of Magnetic Resonance* 178.1 (2006), pp. 42–55. DOI: 10.1016/j.jmr.2005.08.013 (cit. on pp. 55, 62, 76).

[Sun+08] Yongchen Sun, Thomas Böttger, C. W. Thiel, and R. L. Cone. "Magnetic g tensors for the $^4I_{15/2}$ and $^4I_{13/2}$ states of Er^{3+} : Y_2SiO_5". In: *Physical Review B* 77 (8 Feb. 2008), p. 085124. DOI: 10.1103/PhysRevB.77.085124 (cit. on pp. 8, 12, 23, 24, 55, 100).

[Tab+14] Yutaka Tabuchi, Seiichiro Ishino, Toyofumi Ishikawa, Rekishu Yamazaki, Koji Usami, and Yasunobu Nakamura. "Hybridizing Ferromagnetic Magnons and Microwave Photons in the Quantum Limit". In: *Physical Review Letters* 113 (8 Aug. 2014), p. 083603. DOI: 10.1103/PhysRevLett.113.083603 (cit. on p. 4).

[Tak+08] Susumu Takahashi, Ronald Hanson, Johan van Tol, Mark S. Sherwin, and David D. Awschalom. "Quenching Spin Decoherence in Diamond through Spin Bath Polarization". In: *Physical Review Letters* 101 (4 July 2008), p. 047601. DOI: 10.1103/PhysRevLett.101.047601 (cit. on p. 91).

[TC68] Michael Tavis and Frederick W. Cummings. "Exact Solution for an N-Molecule - Radiation-Field Hamiltonian". In: *Physical Review* 170 (2 June 1968), pp. 379–384. DOI: 10.1103/PhysRev.170.379 (cit. on p. 32).

[Tka13] A. Tkalčec. "Cavity QED with Erbium Doped Crystals Coupled to Microwave Resonators". MA thesis. Physikalisches Institut, Karlsruhe Institute of Technology (KIT), 2013 (cit. on p. 75).

[Tka+14] A. Tkalčec, S. Probst, D. Rieger, H. Rotzinger, S. Wünsch, N. Kukharchyk, A. D. Wieck, M. Siegel, A. V. Ustinov, and P. Bushev. "Strong coupling of an Er^{3+}-doped $YAlO_3$ crystal to a superconducting resonator". In: *Physical Review B* 90 (7 Aug. 2014), p. 075112. DOI: 10.1103/PhysRevB.90.075112 (cit. on pp. 4, 64, 65, 107, 113).

[Tog+10] E. Togan, Y. Chu, A. S. Trifonov, L. Jiang, J. Maze, L. Childress, M. V. G. Dutt, A. S. Sorensen, P. R. Hemmer, A. S. Zibrov, and M. D. Lukin. "Quantum entanglement between an optical photon and a solid-state spin qubit". In: *Nature* 466.7307 (Aug. 2010), pp. 730–734. DOI: 10.1038/nature09256 (cit. on p. 3).

[TNM08] Karl Tordrup, Antonio Negretti, and Klaus Mølmer. "Holographic Quantum Computing". In: *Physical Review Letters* 101 (4 July 2008), p. 040501. DOI: 10.1103/PhysRevLett.101.040501 (cit. on p. 3).

[Tyr+12] Alexei M. Tyryshkin, Shinichi Tojo, John J. L. Morton, Helge Riemann, Nikolai V. Abrosimov, Peter Becker, Hans-Joachim Pohl, Thomas Schenkel, Michael L. W. Thewalt, Kohei M. Itoh, and S. A. Lyon. "Electron spin coherence exceeding seconds in high-purity silicon". In: *Nat Mater* 11.2 (Feb. 2012), pp. 143–147. DOI: 10.1038/nmat3182 (cit. on p. 4).

[Usm+10] I. Usmani, M. Afzelius, H. de Riedmatten, and N. Gisin. "Mapping multiple photonic qubits into and out of one solid-state atomic ensemble". In: *Nature Communications* 1 (2010), 12 (7 pp.) DOI: 10.1038/ncomms1010 (cit. on p. 4).

[Ver+09] J. Verdú, H. Zoubi, Ch. Koller, J. Majer, H. Ritsch, and J. Schmiedmayer. "Strong Magnetic Coupling of an Ultracold Gas to a Superconducting Waveguide Cavity". In: *Physical Review Letters* 103 (4 July 2009), p. 043603. DOI: 10.1103/PhysRevLett.103.043603 (cit. on p. 3).

[Wan+09] H. Wang, M. Hofheinz, J. Wenner, M. Ansmann, R. C. Bialczak, M. Lenander, Erik Lucero, M. Neeley, A. D. O'Connell, D. Sank, M. Weides, A. N. Cleland, and John M. Martinis. "Improving the coherence time of superconducting coplanar resonators". In: *Applied Physics Letters* 95, 233508 (2009), p. 233508. DOI: 10.1063/1.3273372 (cit. on p. 46).

[Wis+14] I. Wisby, S. E. de Graaf, R. Gwilliam, A. Adamyan, S. E. Kubatkin, P. J. Meeson, A. Ya. Tzalenchuk, and T. Lindström. "Coupling of a locally implanted rare-earth ion ensemble to a superconducting micro-resonator". In: *Applied Physics Letters* 105.10, 102601 (2014), pages. DOI: 10.1063/1.4894455 (cit. on p. 68).

[Wu+10] Hua Wu, Richard E. George, Janus H. Wesenberg, Klaus Mølmer, David I. Schuster, Robert J. Schoelkopf, Kohei M. Itoh, Arzhang Ardavan, John J. L. Morton, and G. Andrew D. Briggs. "Storage of Multiple Coherent Microwave Excitations in an Electron Spin Ensemble". In: *Physical Review Letters* 105 (14 Sept. 2010), p. 140503. DOI: 10.1103/PhysRevLett.105.140503 (cit. on pp. 3, 94, 97).

[Wue+11] S. Wuensch, G. Hammer, T. Kappler, F. Geuppert, and M. Siegel. "Investigation and Optimization of LEKID Coupling Structures and Multi-Pixel Arrays at 4.2 K". In: *Applied Superconductivity, IEEE Transactions on* 21.3 (June 2011), pp. 752–755. DOI: 10.1109/TASC.2010.2090445 (cit. on pp. 30, 46, 73).

[Yu+02] Yang Yu, Siyuan Han, Xi Chu, Shih-I Chu, and Zhen Wang. "Coherent Temporal Oscillations of Macroscopic Quantum States in a Josephson Junction". In: *Science* 296.5569 (2002), pp. 889–892. DOI: 10.1126/science.1069452 (cit. on p. 28).

[Zag+09] A. M. Zagoskin, A. L. Rakhmanov, Sergey Savel'ev, and Franco Nori. "Quantum metamaterials: Electromagnetic waves in Josephson qubit lines". In: *Physica Status Solidi (b)* 246.5 (2009), pp. 955–960. DOI: 10.1002/pssb.200881568 (cit. on pp. 111, 117).

[Zha+14] Xufeng Zhang, Chang-Ling Zou, Liang Jiang, and Hong X. Tang. "Strongly Coupled Magnons and Cavity Microwave Photons". In: *Physical Review Letters* 113 (15 Oct. 2014), p. 156401. DOI: 10.1103/PhysRevLett.113.156401 (cit. on p. 4).

[Zhu+11] Xiaobo Zhu, Shiro Saito, Alexander Kemp, Kosuke Kakuyanagi, Shin-ichi Karimoto, Hayato Nakano, William J. Munro, Yasuhiro Tokura, Mark S. Everitt, Kae Nemoto, Makoto Kasu, Norikazu Mizuochi, and Kouichi Semba. "Coherent coupling of a superconducting flux qubit to an electron spin ensemble in diamond". In: *Nature* 478.7368 (Oct. 2011), pp. 221–224. DOI: 10.1038/nature10462 (cit. on pp. 3, 28, 64).

[ZAU06] Alexander P. Zhuravel, Steven M. Anlage, and Alexey V. Ustinov. "Measurement of local reactive and resistive photoresponse of a superconducting microwave device". In: *Applied Physics Letters* 88.21, 212503 (2006), pages. DOI: 10.1063/1.2205726 (cit. on p. 56).

[Zhu+12] Alexander P. Zhuravel, Cihan Kurter, Alexey V. Ustinov, and Steven M. Anlage. "Unconventional rf photoresponse from a superconducting spiral resonator". In: *Physical Review B* 85 (13 Apr. 2012), p. 134535. DOI: 10.1103/PhysRevB.85.134535 (cit. on p. 56).

[ZZB10] James F. Ziegler, M.D. Ziegler, and J.P. Biersack. "{SRIM} - The stopping and range of ions in matter (2010)". In: *Nuclear Instruments and Methods in Physics Research Section B: Beam Interactions with Materials and Atoms* 268.11-12 (2010). 19th International Conference on Ion Beam Analysis, pp. 1818–1823. DOI: 10.1016/j.nimb.2010.02.091 (cit. on p. 69).

List of publications

- S. Probst, H. Rotzinger, A. V. Ustinov, P. A. Bushev. ''Microwave multimode memory with an erbium spin ensemble''. In *Phys. Rev. B* 92 (2015), p. 014421, DOI: 10.1103/PhysRevB.92.014421.

- M. Goryachev, W. G. Farr, N. C. Carvalho, D. L. Creedon, J.-M. Le Floch, S. Probst, P. Bushev, M. E. Tobar. ''Discovery of iron group impurity ion spin states in single crystal Y_2SiO_5 with strong coupling to whispering gallery photons''. In: *Appl. Phys. Lett.* 106 (2015), p. 232401, DOI: 10.1063/1.4922376.

- S. Probst, F. B. Song, P. A. Bushev, A. V. Ustinov, M. Weides. ''Efficient and robust analysis of complex scattering data under noise in microwave resonators''. In: *Rev. Sci. Instrum.* 86 (2015), p. 024706, DOI: 10.1063/1.4907935.

- J. Braumüller, J. Cramer, S. Schlör, H. Rotzinger, L. Radtke, A. Lukashenko, P. Yang, S. T. Skacel, S. Probst, M. Marthaler, L. Guo, A. V. Ustinov, M. Weides. ''Multiphoton dressing of an anharmonic superconducting many-level quantum circuit''. In *Phys. Rev. B* 91 (2015), p. 054523, DOI: 10.1103/PhysRevB.91.054523.

- S. Probst, N. Kukharchyk, H. Rotzinger, A. Tkalčec, S. Wünsch, A. D. Wieck, M. Siegel, A. V. Ustinov, P. A. Bushev. ''Hybrid quantum circuit with implanted erbium ions''. In *Appl. Phys. Lett.* 105 (2014), p. 162404, DOI: 10.1063/1.4898696.

- H. Rotzinger, S. T. Skacel, M. Pfirrmann, J. N. Voss, J. Münzberg, S. Probst, P. Bushev, M. P. Weides, A. V. Ustinov, J. E. Mooij. ''Aluminium-oxide wires for superconducting high kinetic inductance circuits''. In: *arXiv:1408.4347* (2014).

- S. Probst, A. Tkalcec, H. Rotzinger, D. Rieger, J-M. Le Floch, M. Goryachev, M. E. Tobar, A. V. Ustinov, P. A. Bushev. ''Three-dimensional cavity quantum electrodynamics with a rare-earth spin ensemble''. In: *Phys. Rev. B* 90 (2014), p. 100404(R), DOI: 10.1103/PhysRevB.90.100404.

- N. Kukharchyk, S. Pal, J. Rödiger, A. Ludwig, S. Probst, A. V. Ustinov, P. Bushev and A. D. Wieck. ''Photoluminescence of focused ion beam implanted Er^{3+}:Y_2SiO_5 crystals''. In: *Phys. Status Solidi RRL* (2014), DOI: 10.1002/pssr.201409304.

- A. Tkalčec, S. Probst, D. Rieger, H. Rotzinger, S. Wünsch, N. Kukharchyk, A. D. Wieck, M. Siegel, A. V. Ustinov, and P. Bushev. ''Strong coupling of an Er^{3+}-doped $YAlO_3$ crystal to a superconducting resonator''. In: *Phys. Rev. B* 90 (2014), p. 075112, DOI: 10.1103/PhysRevB.90.075112.

- S. Probst, H. Rotzinger, S. Wünsch, P. Jung, M. Jerger, M. Siegel, A. V. Ustinov, and P. A. Bushev. ''Anisotropic Rare-Earth Spin Ensemble Strongly Coupled to a Superconducting Resonator''. In: *Phys. Rev. Lett.* 110 (2013), p. 157001, DOI: 10.1103/PhysRevLett.110.157001.

- A. Jesche, F. Nitsche, S. Probst, Th. Doert, P. Müller, and M. Ruck. ''Anisotropic electrical resistivity of LaFeAsO: Evidence for electronic nematicity''. In: *Phys. Rev. B* 86 (2012), p. 134511, DOI: 10.1103/PhysRevB.86.134511.

Acknowledgements

The results presented in this thesis would not have been possible without support of many people.

First, I would like to thank Prof. Dr. Alexey V. Ustinov for giving me the opportunity to work on this great topic. He provided me with a lot of freedom to pursue my own ideas and gave me the possibility to visit many conferences.

Next, I am very grateful to Prof. Dr. Pavel A. Bushev who has been supervising me during this thesis. We were a good team and had a lot of stimulating discussions about physics and the rest of the world.

Special thanks go to Dr. Hannes Rotzinger. He was always available for discussing current problems with the experiment, and I appreciate his many clever solutions. I would like to thank Andrej Tkalčec and Daniel Rieger for the good and productive time we had during their Master thesis. The project benefited from a great collaboration within the QUIMP project. I gratefully acknowledge support by Dr. Stefan Wünsch who fabricated the excellent superconducting resonators used in this thesis. Furthermore, I would like to thank Prof. Dr. Andreas D. Wieck and Nadezhda Kuckharchyk from the Ruhr University Bochum for the good collaboration on the implanted spins project. Many thanks go to Dr. Martin Weides and Dr. Fengbin Song for interesting discussions on resonators and the good collaboration. I want to thank Dr. Olexandr Lukasheko for the technical support and many helpful advice. He taught me a lot about cryostats.

Special thanks go to Dr. Susanne Butz and Dr. Jürgen Lisenfeld for the excellent proofreading of the thesis. I am happy that I had the opportunity to work in such a stimulating and creative environment with many good colleges and friends. Many thanks go Dr. Susanne Butz and Jochen Braumüller for the good time we had. Also, I always enjoyed the many crazy discussions with Sebastian Skacel and I hope that we can continue our project.
I would like to acknowledge the great support by the mechanical and electronic workshop. Without their quick and professional work, the experimental setup would not have been realizable. Many thanks also go to the secretaries office and Lars Behrens.

I would like to thank my parents and my brother Matthias for their endless support throughout my life. Finally, special thanks go to Anne and my friends. Thank you for the great time we had together and for your support.

I gratefully acknowledge financial support by the Landesgraduiertenförderung of Baden-Württemberg, and the BMBF program "Quantum communications" through the project QUIMP.

Appendix

Nomenclature

Symbol	Meaning	Comment
υ	coupling strength	in units of 2πHz
$\Gamma_2^\star$	inhomogeneous linewidth	in units of 2πHz ‡
γ	homogeneous linewidth	in units of 2πHz ‡
κ	resonator linewidth	in units of 2πHz ‡
$\Delta f_{3\mathrm{dB}}$	resonator bandwidth	$\Delta f_{3\mathrm{dB}}=2\kappa$
Q_c	coupling quality factor	
Q_i	intrinsic quality factor	
Q_t, Q_r	total quality factor	$1/Q_t = \sum_j 1/Q_j$
$\delta_\times$	tunnel coupling energy in a flux qubit	
Δ	frequency detuning	
ϵ	electric moment	
$\boldsymbol{\mu}$	magnetic moment	
$\mathbf{g}$	**g**-Tensor	
g, g	g-factor	
L	total orbital quantum number	L=S,P,D,F,G,H,$I\cdots$ or L=0,1,2,3,4,5,6$\cdots$
S	total spin quantum number	
J	total angular momentum quantum number	
χ	susceptibility	
ξ	filling factor	

‡ half width at half maximum (HWHM)

Magnetic field angles

The orientations of the magnetic field with respect to the crystal frame are given by the angles ϕ and θ.

- ϕ: angle between x axis and xy plane (for YSO: D1=x, xy-plane=D1D2-plane)
- θ: angular deviation from the z axis (for YSO: b)

Pulse calibration

The amplitude of the microwave pulses is controlled by a microwave mixer (Marki M8D420LS). Figure 1 shows the measured pulse energy vs. the voltage amplitude at the IF port of the mixer. As one can see, the mixer's transmission is not a linear function of the IF voltage.

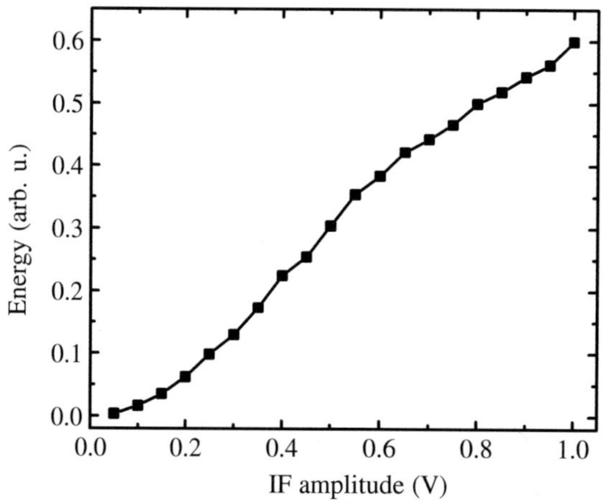

Figure 1: IF amplitude vs. relative pulse energy.

Experimental Condensed Matter Physics
(ISSN 2191-9925)

Herausgeber
Physikalisches Institut

Prof. Dr. Hilbert von Löhneysen
Prof. Dr. Alexey Ustinov
Prof. Dr. Georg Weiß
Prof. Dr. Wulf Wulfhekel

Die Bände sind unter www.ksp.kit.edu als PDF frei verfügbar
oder als Druckausgabe bestellbar.

Band 1 Alexey Feofanov
Experiments on flux qubits with pi-shifters. 2011
ISBN 978-3-86644-644-1

Band 2 Stefan Schmaus
Spintronics with individual metal-organic molecules. 2011
ISBN 978-3-86644-649-6

Band 3 Marc Müller
Elektrischer Leitwert von magnetostriktiven Dy-Nanokontakten. 2011
ISBN 978-3-86644-726-4

Band 4 Torben Peichl
Einfluss mechanischer Deformation auf atomare Tunnelsysteme – untersucht mit Josephson Phasen-Qubits. 2012
ISBN 978-3-86644-837-7

Band 5 Dominik Stöffler
Herstellung dünner metallischer Brücken durch Elektromigration und Charakterisierung mit Rastersondentechniken. 2012
ISBN 978-3-86644-843-8

Band 6 Tihomir Tomanic
Untersuchung des elektronischen Oberflächenzustands von Ag-Inseln auf supraleitendem Niob (110). 2012
ISBN 978-3-86644-898-8

Band 7 Lukas Gerhard
Magnetoelectric coupling at metal surfaces. 2013
ISBN 978-3-7315-0063-6

Experimental Condensed Matter Physics
(ISSN 2191-9925)

Band 8 Kirill Fedorov
Fluxon readout for superconducting flux qubits. 2013
ISBN 978-3-7315-0067-4

Band 9 Jochen Zimmer
Cooper pair transport in arrays of Josephson junctions. 2014
ISBN 978-3-7315-0130-5

Band 10 Oliver Berg
Elektrischer Transport durch Nanokontakte von Selten-Erd-Metallen. 2014
ISBN 978-3-7315-0209-8

Band 11 Grigorij Jur'evic Grabovskij
Investigation of coherent microscopic defects inside the tunneling barrier of a Josephson junction. 2014
ISBN 978-3-7315-0210-4

Band 12 Cornelius Thiele
STM Characterization of Phenylene-Ethynylene Oligomers on Au(111) and their Integration into Carbon Nanotube Nanogaps. 2014
ISBN 978-3-7315-0235-7

Band 13 Michael Peter Schackert
Scanning Tunneling Spectroscopy on Electron-Boson Interactions in Superconductors. 2014
ISBN 978-3-7315-0238-8

Band 14 Susanne Butz
One-Dimensional Tunable Josephson Metamaterials. 2014
ISBN 978-3-7315-0271-5

Band 15 Philipp Jung
Nonlinear Effects in Superconducting Quantum Interference Meta-Atoms. 2014
ISBN 978-3-7315-0294-4

Band 16 Sebastian Probst
Hybrid quantum system based on rare earth doped crystals. 2016
ISBN 978-3-7315-0345-3